T0320643

Balanced
Sustainability in
a Changing
World

Exploring Complexity

ISSN: 2382-5901

For four centuries our sciences have progressed by looking at its objects of study in a reductionist manner. In contrast complexity science, that has been evolving during the last 30–40 years, seeks to look at its objects of study from the bottom up, seeing them as systems of interacting elements that form, change, and evolve over time. Complexity therefore is not so much a subject of research as a way of looking at systems. It is inherently interdisciplinary, meaning that it gets its problems from the real non-disciplinary world and its energy and ideas from all fields of science, at the same time affecting each of these fields.

The purpose of this series on complexity science is to provide insights in the development of the science and its applications, the contexts within which it evolved and evolves, the main players in the field and the influence it has on other sciences.

For the complete list of volumes in this series, please visit www.worldscientific.com/series/ec

Exploring Complexity – Volume 10

Balanced Sustainability in a Changing World

Editors

Ernst Pöppel
Ludwig Maximilian University Munich, Germany

Maria Reinisch
Federation of German Scientists

 **World Scientific**

NEW JERSEY · LONDON · SINGAPORE · BEIJING · SHANGHAI · HONG KONG · TAIPEI · CHENNAI · TOKYO

Published by

World Scientific Publishing Co. Pte. Ltd.

5 Toh Tuck Link, Singapore 596224

USA office: 27 Warren Street, Suite 401-402, Hackensack, NJ 07601

UK office: 57 Shelton Street, Covent Garden, London WC2H 9HE

British Library Cataloguing-in-Publication Data
A catalogue record for this book is available from the British Library.

Exploring Complexity — Vol. 10
BALANCED SUSTAINABILITY IN A CHANGING WORLD

ISBN 978-981-128-430-4 (hardcover)
ISBN 978-981-128-431-1 (ebook for institutions)
ISBN 978-981-128-432-8 (ebook for individuals)

For any available supplementary material, please visit
https://www.worldscientific.com/worldscibooks/10.1142/13624#t=suppl

Printed in Singapore

Foreword

Pragmatic suggestions by the young generation for sustainability in a balance

In autumn 2023, the Federation of German Scientists together with the Institute of Medical Psychology, Ludwig Maximilian University, organized an interdisciplinary and international "autumn school" in which 30 young researchers and students participated. We used a new format of academic teaching that allowed to bring together students from different disciplines and different cultural backgrounds. The idea of the invitation was that the participants make pragmatic suggestions for the solution of imminent global and local problems. Participants came from these countries: Brazil, China, Germany, Italy, Nepal, Netherlands, Norway, Russia, Taiwan, Turkey, UK, USA. This distribution of representatives from different countries shall also reflect our motto: "scientists are natural ambassadors", and this applies to members of all generations, young and old.

The autumn school — being indeed an "academic experiment" — started with a series of ten online lectures that were always delivered on weekends. These times were chosen in order not to interfere too much with other academic duties of the participants in the different countries. These lectures by international leading scientists and prominent experts from different academic fields were meant to provide background knowledge for "balanced sustainability in a changing world". Only some key words are given here. Topics selected in the first block of lectures focused on climate change and biodiversity. For the anthropogenic climate change (which is taken for granted), it was questioned whether the international laws are sufficient for control. This has been seriously questioned. It was then analysed how the crises of biodiversity, climate change and pandemics are interrelated. These crises cannot and should not be treated independently. A very specific and surprising suggestion was made for a sustainable solution of CO_2

reduction; this may be possible with the help of the Cyanobacterium "Spirulina", an ancient planetary organism. In the second block, lectures were given on social and personal challenges in a changing world: How should "smart cities" be organized in the future, and how a successful urban transformation can be managed. Then it was explained how a neuropolitical approach to humanization can help to overcome polarization in countries. Obviously, different political systems have to work together. This aspect of a necessary social togetherness was also addressed in the third block with an example from Nepal and beyond. It was stressed that an energy transition for a world of renewable energies requires individual involvement made possible by educating the young and the elderly. Then in a hierarchical system integrating concepts of brain mechanisms, evolutionary principles and ethical constraints, a dynamic equilibrium for sustainability was described. Finally, in a fourth block, questions of peace and intercultural spirit were evaluated. How can we reach and maintain equanimity in a changing world? How can peace be obtained? We seem to be confronted with a frightening future believing in good old times. What could be the role of utopias or dystopias to get a deeper understanding of the changing world in the present? All these lectures were followed by intense discussions with the students, and their questions reflected the interdisciplinary background of the participants.

During this first phase of the "autumn school" with the lectures, teams of three participants from different countries and with different academic background were formed. The task for the small groups was to develop during the next four weeks, the second phase of the "autumn school", practical solutions for imminent problems that were addressed during the lectures, but the participants were free to choose their own topics which they considered to be relevant or interesting. The participants had regular online meetings with some organizational support from the office of the Federation of German Scientists, but with no interference in their creative activity when they were exploring the complexity of the global and local challenges in a changing world. (The organizers are very grateful for financial support by the Andrea von Braun Stiftung, Munich, Germany, with its interdisciplinary portfolio also being dedicated to environmental issues.)

The results of the second phase were presented by the participants in early December 2023, and the results were critically discussed between the participants and several expert lecturers from the first phase. These discussions and further reflections about the presented projects started the third phase of the "autumn school". The task was now to document in written form what has been achieved. Partly new groups were formed, and during the next months, manuscripts were written in an interactive mode. These manuscripts were critically reviewed between members of the autumn school, and more information was gathered about the different topics when necessary. Finally, six manuscripts were submitted to the organizers which are presented in this book. It is important to note that everything presented is the work of the "young generation" only. They could enjoy their independence and creativity, and they could also experience their responsibility as global citizens. The young generation grasped the opportunity to develop their own ideas, they did not have the feeling to be exploited but explored trajectories of pragmatic solutions, either triggered by the lectures they had heard or on the basis of their own interest. This "academic experiment" was not only successful with respect to pragmatic solutions developed, but also from a political point of view: The participants came from different countries with different political systems, with different historical background, different belief systems, and although they never met in person, modern technology allowed their mutual intellectual interaction to solve problems in a joint task. They became friends "at a distance" and between them trust was growing. Indeed, "scientists are natural ambassadors", in particular with a task for achieving a common goal.

Editors
Ernst Pöppel and Maria Reinisch
Institute of Medical Psychology, Ludwig Maximilian University Munich, Germany, Federation of German Scientists (VDW), Berlin, Germany

Contents

1 Urban nature(s): Conceptualization and interculturality

Elvira Barkhatova
Immanuel Kant Baltic Federal University (IKBFU), Russia

Andrew Turner Poeppel
Centre for Development and the Environment University of Oslo, Norway

Haiming Yang
School of Psychological and Cognitive Sciences, Peking University, China

1.1 Abstract

Global ecological crises place significant strain on cities at a time of accelerating urbanization. 55% of the world's population currently live in urban areas, a proportion that is expected to increase to 68% by 2050. These areas continue to experience rapid development and confront ecological challenges including climate change, biodiversity loss, and environmental degradation. As a result, an increasing number of urban policymakers have turned to sustainability initiatives, and environmental movements have called for greater involvement in decision-making processes at the local and political levels. However, strategies to reduce the ecological footprint of cities and establish "green" development practices remain vigorously debated among policymakers, city dwellers, and researchers. Given the need to implement socially inclusive and environmentally just frameworks of collective action in the built environment, strategies centered on citizen engagement are essential. However, understandings of "nature" and "sustainability" can vary significantly across social and cultural boundaries, so citizen engagement cannot be viewed monolithically. How can effective communication around urgent socio-ecological issues occur within the domain of urban policymaking? Here we examine how intercultural communication impacts the perception of socially inclusive and ecologically sustainable urban environments. We offer two distinct disciplinary approaches to examine the complexity of urban environmental identity and point to the dynamic interaction between personal and collective relationships to urban ecologies. The dialogue outlines how socio-cultural perspectives can improve our understanding of the connections between city dwellers and urban nature(s). It emphasizes the need for further research on intercultural engagement in urban policymaking and argues that the complexity of urban environmental identity must be taken into consideration to promote effective, cross-cultural communication. Focusing on the cross-cultural dialogue of the 2022 Autumn School for the Young VDW, the paper provides two perspectives on the issues of interdisciplinarity and interculturality in urban policy making and planning.

1.1.1 *Intercultural cities: Urban development and the ecological crisis*

Among the most significant social and technical challenges of the twenty-first century is the establishment of environmentally ethical strategies to address increasing urbanization and global environmental crises. Cities are home to over 4.4 billion inhabitants, and current projections indicate that roughly 68% of the world's population will live in urban areas by 2050 (UN 2018). The post-1950 acceleration of development activities in the "urban sphere" has brought attention to unsustainable practices of planning and design in the built environment. Additionally, public awareness of the climate crisis and political pressure by popular movements centered on justice and ecology have led an increasing number of policy-makers to pursue development strategies centered on "urban sustainability". These strategies vary significantly in their objectives, and despite their varying degrees of success, they generally seek to decrease the adverse impacts of urbanization and development practices (Alberti and Susskind 1996). City planners have also put forward development plans to promote climate resilience in numerous geographic contexts, with the increased use of policy vocabulary around civic engagement and democratic governance (Healey 2008). While this emphasis on public participation can be identified in contemporary urban planning discourses, the implementation of sustainability frameworks in the built environment remains highly contested, both in the academy and among environmental movements.

The paper expands on the dialogue that occurred in the 2022 Autumn School to examine how interdisciplinary and intercultural communication impacts the perception of socially inclusive and environmentally sustainable cities. While research on the *intercultural city* already draws from several disciplinary backgrounds and intellectual traditions, its connection to urban sustainability remains undertheorized. The paper first provides an overview of the theoretical and conceptual frameworks that may shape the discussion of urban environments. This includes an interdisciplinary reflection drawing from neuroscience, outlining a neural model for understanding the built environment. Thereafter, the paper discusses how socio-cultural perspectives can illustrate the dynamic relationships that exist between urban citizens and urban nature(s). The analysis draws from the use of comparative practice in urban studies (Robinson 2016, 2022) to generate new conceptual frameworks for "the intercultural city" (White 2018). This approach points to differentiated understandings of sustainability and environmental identity, as well as the path towards effective cross-cultural communication to promote citizen engagement.

Given the importance of social inclusion in the effort to promote public engagement on climate change, the incorporation of intercultural perspectives into contemporary planning discourses is essential. Building on intercultural and civic-oriented approaches to urban planning and design may therefore be one effective avenue to explore the diversity of environmental perspectives and biospheric values in cities.

1.1.2 *Conceptual metaphors of the city*

Throughout history, there have been various ways of conceptualizing cities (and the built environment). These conceptual renderings have been conditioned by the existing economic and political paradigms in given societies. The discipline of urban studies has seen the development of conceptual metaphors (CM) such as THE CITY IS A MECHANISM (MACHINE) and THE CITY IS A LIVING ORGANISM. The first of these conceptual metaphors is, arguably, not fully applicable in the face of contemporary research on the living components of the city itself. These findings point to urban environments as complex ecological systems, and not a simple assembly of constituent parts. A city can therefore be depicted as an emergent structure which is larger than the sum of its parts. Although the second metaphor resolves the problem of "mechanisticity", it also produces new conceptual tension, for instance, the asymmetry of functions and, consequently, the disbalance of esthetic value. The "trite" or conventional metaphors such as "the city center is its heart", or "green areas are the lungs of the city" are widely used in discourses surrounding the built environment. However, such physiological comparisons may invite more questions, rather than provide actionable answers. The list of the possible conceptual metaphors is not exhaustive, and numerous frameworks have been proposed as conceptual solutions. These include the city as a form of collective intelligence (Villarreal 2020), the built environment as a socio-ecological system (Moffatt and Kohler 2008), and "neural smart city models" (de Castro and McKenzie 2022), among others.

Throughout the 2022 Autumn School of the Young VDW, titled *Balanced Sustainability in a Changing World*, the Urban Nature(s) research group engaged in an ongoing dialogue surrounding the conceptual metaphors that might aid in the discussion of intercultural and sustainable cities. This dialogue included extensive discussion of the concept of sustainability itself, and its translation into the planning and design of socially just and environmentally ethical cities. The discussions, both internal and external, produced valuable results, particularly when engaging with researchers from diverse disciplinary and cultural backgrounds. This diversity of thought was demonstrated by the wide range of responses to the discussion around conceptual metaphors. Given the value of these responses, the intention is to further explore the conceptual metaphor of THE CITY IS A NEURON, an idea expressed by researcher Chen Zhao. It is worth pointing out that this conceptual metaphor is subject to interpretation, and highlights a new theoretical perspective on the conceptualization of cities. This approach to the conceptualization of urban space brings forward crucial aspects such as connectivity, efficiency of function division, and harmony of organization.

Here the conceptual model drawing from neuroscience will be examined to highlight how theoretical understandings of urban space can be applied to analyze the socio-ecological dynamics of the city. Following this analysis, the paper will transition to a discussion of how these diverse interpretations of urban nature(s) can be incorporated into planning practices and the discourse around sustainable solutions. Applying the Conceptual Metaphor Theory by G. Lakoff and M. Johnson (1980), the structure of the CM consists of a source domain ("neuron") and a target domain ("city"), i.e. the concept of a city is examined in light of potentially informative parallels of the structure of a neuron. The target domain is structured by the frame [Locale by use], which contains such slots as Locale (a stable bounded area), Constituent_parts (salient parts that make up a Locale), Container_possessor (the location that the Locale is a part of), Descriptor (feature or characteristic of the Locale), and Relative_location (a place that a Locale is located with respect to) (FrameNet). In case of a *city*, the slots can be filled in the following way:

Locale	The city
Constituent_parts	Administrative districts
Container_possessor	The Country or Municipality *(Larger entity which the city is part of)*
Descriptor	Inhabited; densely populated; large; incorporated; established by state
Relative_location	Geopolitical location of the city

Fig. 1. Structure of the city frame and its conceptual features.

In case of a *neuron*, the conceptual filling may be viewed as follows:

Locale	The neuron
Constituent_parts	The soma, axons, dendrites
Container_possessor	The brain
Descriptor	Polarized (receptive & communicating)
Relative_location	Upper part of the body

Fig. 2. Structure of the neuron frame and its conceptual features.

How can a comparative analysis of these target domains improve our understanding of the properties of cities? According to *Principles of Neural Science* (2000), "the great diversity of nerve cells — the fundamental units from which the modules of the nervous systems are assembled — is derived from one basic cell plan. Three features of this plan give nerve cells the unique ability to communicate with one another precisely and rapidly over long distances". This ability of a neuron can be seen as analogous to communication between cities. Within a cell, "first, the neuron is polarized, possessing receptive dendrites on one end and communicating axons with synaptic terminals at the other". This feature of a neuron correlates to urban logistics. "Second, the neuron is electrically and chemically excitable. Its cell membrane contains specialized proteins — ion channels and receptors — that permit the influx and efflux of specific inorganic ions, thus creating electrical currents that alter the voltage across the membrane". The idea of the movement of ions can be paralleled to the circulation of people and goods, to import and export, while the changing voltage can correspond to constantly changing economic and political dynamics within the urban area. "Third, the neuron contains proteins and organelles that endow it with specialized secretory properties that allow it to release neurotransmitters at synapses". Here, the inner structure of a nerve cell (proteins and organelles) corresponds to the municipal structures, both physical and administrative, and their function of releasing neurotransmitters is in line with producing certain impacts on the country's development. (Principles of Neural Science). If we extend the metaphor of THE CITY IS A NEURON even further, we may argue that city satellites (towns or areas around larger cities) can be compared, for example, to glial cells, which execute a supporting function.

Therefore, the overall framework for a conceptual metaphor centered on the "city as a neuron" can be presented in the following way:

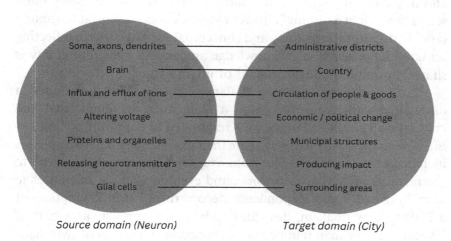

Source domain (Neuron)　　　　　　Target domain (City)

This exercise in alignment between two distinct disciplinary perspectives could stimulate new thinking on the issue of *balanced sustainability* in cities; by providing an "intuition pump", it may incentivize readers to think in novel ways about the organization of the built environment.

1.1.3 *Navigating urban nature(s) and sustainable solutions*

The interdisciplinary dialogue that took place during the 2022 Autumn School on *Balanced Sustainability* highlighted the diverse interpretations of urban environments at a time of accelerating planetary change. The discussions and presentations emphasized how intercultural dialogues can bring new perspectives on a range of socio-ecological issues taking place in urban space. By encouraging responses from a range of cultural (and disciplinary) backgrounds, the meeting brought attention to the question of how cultural identities shape the implementation of genuinely sustainable solutions in the built environment. Contemporary urban studies research has addressed the difficulty of understanding how city dwellers and urban planners relate to complex socio-ecological systems. Childers *et al.* (2014, p. 325) argue that "understanding how cities, as complex adaptive social–biophysical systems, behave seems overly challenging, and moving beyond understanding to identifying and implementing real-world solutions for urban sustainability often seems downright daunting". In order to tackle this "downright daunting task" with cultural sensitivity and conceptual rigor, it is worth reflecting on how the production of knowledge around the intercultural city is shaped by cultural understandings of urban nature(s).

Part of the dialogue on "balanced sustainability" included a discussion with researchers from a wide range of intellectual and cultural backgrounds. Participants responded to the prompt: "*The city is…?*" with the ability to freely share their associations. These responses included associations ranging from *ecosystem* and *mechanical systems* to abstractions such as *home* or *circle*, and even pop-culture references to *mortal engines*. Some respondents deconstructed the assumption of a Nature/City dualism, describing urban environments as a part of "Nature" rather than reinforcing the conceptual division of manmade versus ecological systems. Here an implicit challenge was put to the underlying dualist thinking that so often characterizes discussions of urbanism within the academy. However, this deconstruction of the Nature/City divide was not unanimous, with other respondents comparing built environments to mechanical systems such as vehicles or engines. While not seeking to resolve the tension that emerges from subjective interpretations of conceptual metaphors, the diversity of responses provided an opportunity to examine the benefits of intercultural communication around urban ecological issues.

The concept of the "built environment" has undergone significant change in the decades following the Great Acceleration and the dramatic rise in development activities. Concern over the environmental impact and the inequities of manmade systems has increased with growing awareness of the climate (and biodiversity) crisis. As a result, discussions around the socio-ecological dynamics of cities have received increased attention within professional circles of "urbanists".

Generally speaking, the domain of urban planning has defined the built environment "in contrast to the 'un-built' environment, or the eco-sphere" (Moffatt and Kohler 2008). Within this framework, both the city itself, and its surrounding natural environment can be seen as self-reproducing systems that form loose, nested hierarchies (ibid., p. 249). Recent scholarship combining environmental and urban perspectives has challenged the built/unbuilt dichotomy, encouraging more careful consideration of the *more-than-human city* and relational ideas of place (Franklin 2017, Robertson 2018). Therefore, it is necessary to acknowledge that the relation between the two is by no means historically fixed. "Cultural perceptions of what is meant by society and nature are also changing..." (ibid.), arguably even more so as a result of accelerating urbanization, climate change, and the planetary-scale changes associated with the Post-Holocene. These historical and material developments, while certainly more daunting than the conceptualization of the city, do not invalidate theoretical discussions of urban space. Geographer and urbanist Matthew Gandy (2002, p. 2) argues that "nature has a social and cultural history that has enriched countless dimensions of the urban experience. The design, use, and meaning of urban space involve the transformation of nature into a new synthesis." Responding to Gandy's statement from the perspective of urban political ecology, Heynen *et al.* (2006, p. 5) argue that "understanding the politicized and uneven nature of this urban synthesis should be the main task." Following the productive dialogue on "balanced sustainability", we contend that this understanding of urban experience can be enriched by intercultural dialogue.

Intercultural communication in the domain of "urbanism" re-centers the conversation on the *social* experience of the built environment, as well as the *distinct interpretations of urban nature(s)*. It also emphasizes that these diverse understandings existing among city dwellers are critical sources of knowledge needed to address the unevenly distributed impact of development practices. Although residents of urban areas often go unrecognized as "authorities" on the consequences of urban planning and development, firsthand experience with the impact of changes at street level is an essential source of information for policy makers. City dwellers who continuously adapt to the changes brought on by the urban metabolism arguably have a nuanced (if typically implicit) understanding of the dynamic relationships that form between communities and the material environment. These relationships are shaped by the uneven nature of development practices (Heynen *et al.* 2006), whether it be in the case of establishing "green" public space or innovative sustainable design solutions. Therefore, adhering to the principles of sustainability or green design may have entirely separate meanings to different communities residing in the same city. At the neighborhood or even street level, these dynamic relationships and understandings of urban sustainability may not correspond with the overarching visions of urban planners and designers. As a result, a knowledge gap is formed between those engaging in the professional disciplines that shape the built environment and those city dwellers who must experience the impacts of development practices.

This ethical challenge reveals why intercultural dialogue, particularly at the city planning level, is necessary to address the divisions between urbanists and city dwellers. Note that research on intercultural cities has emphasized this ethical concern. White (2018), for example, argues that the intercultural cities framework goes beyond recognizing the diversity advantage. It "[attempts] to go from a model of peaceful coexistence to one that permits meaningful sustained interactions between groups" (ibid., p. 28). Navigating the diverse understandings of urbanity and sustainability that exist in intercultural cities therefore requires more careful consideration of the rifts that form as a result of development practices. In order to create socially inclusive frameworks that address pressing ecological challenges, the connection between environmental values and socio-cultural experience of the built environment must be addressed. This includes the recognition that city dwellers are key knowledge producers around urban environmental change as well as the impacts of planning or design paradigms. In the context of *balanced sustainability* and the VDW's Autumn School, it follows that discussions of sustainable solutions ought to consider the question: *what* exactly is being sustained, and for *whom?* Given that accelerating urbanization points towards increased cultural diversity in the built environment, the rethinking of sustainable planning and design solutions must respond to social and cultural needs. Intercultural dialogue, matched with sustained engagement with civic and activist groups, may be one avenue to avoid urban development clichés and promote more inclusive planning frameworks.

The topic of "balanced sustainability" also initiated a discussion around the limits and pitfalls of top-down approaches to planning and development. Sustainability has certainly become a marketable term in urban discourse, as evidenced by the international *Green City Awards*. However, cities lauded for their ambitious environmental targets and sustainability initiatives may point to planning strategies that were developed without an emphasis on social, cultural, or civic engagement. To prevent the consequences of uneven and ecologically destructive development practices, it is therefore necessary to consider how bottom-up approaches can re-emphasize social needs. This conclusion points to the requirement that organizing efforts connect planners, architects, designers, and city dwellers to navigate the multiplicity of social and ecological relations. It not only broadens the arena of participation around sustainable cities, but also offers a path to understand how development practices relate to conceptual understandings of urban space. This approach is in line with Robinson's (2022) vision of "generating concepts of 'the urban' through comparative practice." While naturally limited in scope, the discussions that took place on intercultural cities and "balanced sustainability" underscored the path towards new forms of socio-cultural engagement in urban policy making and planning. Dialogue between city dwellers and urbanists on ecological issues, matched by an intercultural perspective, may aid in the development of new frameworks for citizen engagement that promote urban environmental justice. Public participation on urban ecological issues is crucial, and an intercultural approach may produce innovative frameworks for citizen engagement at the city level that promote collective action and ethical concern for urban nature(s).

References

Alberti, M. and Susskind, L. "Managing urban sustainability: An introduction to the special issue." *Environmental Impact Assessment Review* 16, nos. 4–6 (1996): 213–221.

Childers, D. L., Pickett, S. T. A., Grove, J. M., Ogden, L. and Whitmer, A. "Advancing urban sustainability theory and action: Challenges and opportunities." *Landscape and Urban Planning* 125 (2014): 320–328.

de Castro, R. and McKenzie, R. "Introducing the Neural Smart City Model." *Cities Today*, January 10, 2022.

Franklin, A. "The more-than-human city." *The Sociological Review* 65, no. 2 (2017): 202–217.

Gandy, M. *Concrete and Clay: Reworking Nature in New York City.* Cambridge, MA/London: The MIT Press (2002).

Healey, P. "Civic engagement, spatial planning and democracy as a way of life." *Planning Theory & Practice* 9, no. 3 (2008): 379–414. https://doi.org/10.1080/14649350802277092.

Heynen, N., Kaika, M. and Swyngedouw, E. "In the nature of cities." *Urban political ecology and the politics of urban metabolism.* London, New York (2006).

Kandel, E. R., Schwartz, J. H., Jessell, T. M., Siegelbaum, S., Hudspeth, A. J. and Mack, S. eds. *Principles of Neural Science* Vol. 4. New York: McGraw-hill, 2000.

Lakoff, G. and Johnson, M. "The metaphorical structure of the human conceptual system." *Cognitive Science* 4, no. 2 (1980): 195–208.

Moffatt, S. and Kohler, N. "Conceptualizing the built environment as a social–ecological system." *Building Research & Information* 36, no. 3 (2008): 248–268.

Robertson, S. A. "Rethinking relational ideas of place in more-than-human cities." *Geography Compass* 12, no. 4 (2018): e12367.

Robinson, J. "Thinking cities through elsewhere: Comparative tactics for a more global urban studies." *Progress in Human Geography* 40, no. 1 (2016): 3–29.

Robinson, J. "Introduction: Generating concepts of 'the urban' through comparative practice." *Urban Studies* 59, no. 8 (2022): 1521–1535.

Villarreal, O. *Biocoenosis Nest: A City Woven by Collective Intelligences.* London, UK: Bartlett School of Architecture, 2020.

White, B. W. "What is an intercultural city and how does it work?" *Intercultural Cities: Policy and Practice for a New Era* (2018): 21–54.

_____ "68% Of the World Population Projected to Live in Urban Areas by 2050." UN Department of Economic and Social Affairs, May 16, 2018.

FrameNet Database. URL: https://framenet.icsi.berkeley.edu/fndrupal/framenet_search.

2 Rethinking of structural, scientific, and spiritual solutions for our relationship with the environment based on indigenous knowledge

Amanda Vicentini
Department for Theology, Pontifical Catholic University of Paraná, Curitiba, Brazil

Philipp Geyer
Department for Political Science,
Goethe Universität Frankfurt am Main,
Frankfurt a.M. Germany

Sophia Reinisch
Department for Geography, Ludwig Maximilian University of Munich, Munich, Germany

Yingrui Luo
School of Psychological and Cognitive Sciences, Peking University, Beijing, China

2.1 Abstract

In the process of enriching our scientific practice, our political discourse and strengthening our relationship with the planet and all beings, the objective of this article is to shed light on some practices and wisdom of indigenous peoples. Understanding that the knowledge of these original peoples, who maintain a relationship of familiarity with all of nature, must be a resource that needs to gain strength and legitimacy. There is a need for a paradigm shift in the modus operandi of Western

15

culture and science, which can, step by step, open up to different ways of understanding and inhabiting the world.

In this effort, we want to work with three main axes. First, we will analyze the place of indigenous knowledge in the current scientific community, and how this knowledge can be crucial in the process of global warming. The second part will reflect on shamanic forms of spirituality, focusing on the use of an ancient drink, ayahuasca, and how the experience with this drink can help individuals to better connect with themselves, with their group and the entire planet. We end this article by showing how the implementation of indigenous practices in urban infrastructures has proven to be efficient and sustainable alternatives, and increases the quality of life in cities.

2.1.1 Introduction

In a world in transformation, science must open its eyes to alternative forms of understanding the whole. In a society that has lost its enchantment, ancestral forms of relationality with the world have emerged and become popular in an attempt to rescue this lost bond. In this way, seeing the fragility of an actual model of science that does not meet many demands and challenges of our time, our task is an operation of epistemological rescue.

To rescue knowledge, epistemologies and practices which are not sufficiently implemented in traditional science parameters should find more audience in the broader public. It is the rescue of knowledge, when epistemologies and practices are explored which did not fit into the traditional science parameters and formulas, and were relegated and rejected in a mission of colonization, indoctrination and rooting of an exclusively Christocentric, European, positivistic and rational cosmovision, which influenced our relationship with nature and the way we are perceiving and feeling the world.

Our intention with this work is to enhance the reflection and open paths to the discussion about the way of living by indigenous people and how the cosmosensations,[1] can help us to re-encounter the ideal

---

[1] The sense of "cosmosensation" was created by the feminist researcher and associate professor of sociology at Stony Brook University, the Nigerian OyèrónkéOyěwùmí. Criticizing the concept of worldview, she coined this

situation of human living in harmony with nature and the whole cosmos.

Our paper will be separated into three parts, all with a different focus on indigenous knowledge. First, we want to examine the place of indigenous knowledge in the current scientific community. There is much agreement that indigenous knowledge is critical for the adaptation process to the climate crisis. Accordingly, more attention should be paid to indigenous knowledge. We will analyze indigenous adaptation options that could benefit the local region, larger populations, and which can potentially be applied supraregional.

Following this, we will discuss how ancestral forms of spirituality can help us to live in a more sustainable, integral and harmonious world. We will focus here on a specific shamanic practice, the consumption of ayahuasca, an ancient indigenous drink that has been crossing the borders of the forest and gaining interest in various fields of knowledge. This practice is based on two main principles, learning and healing, showing a fantastic ability to promote enduring effect on individual, group well-being and society, even providing increased ecological and planetary care.

In our next part, we want to set the focus on how we can adapt urban structures to a more sustainable and healthier lifestyle by taking ecosystem services and indigenous knowledge into consideration. Bringing nature back into the city, adjusting the infrastructure and lifestyle to the local climate, learning about the domestic biodiversity and restoring natural safety systems like wetlands and floodplans are possibilities that can help to raise the quality of life in cities.

In our last part, we want to focus on one particular way to bring nature back in urban areas. In a more application-oriented matter we focus on urban gardening, a challenging but also promising way. We want to combine theoretical and practical concepts. And we especially think that our mindset towards our current economic and ecological systems is what needs to be changed and that is where indigenous communities are way ahead of us.

---

term for one to understand a "world-sense" as opposed to a "worldview" or cosmovision used in the West.

2.1.2 *Indigenous knowledge. Their wisdom, our responsibility*

Communication and respect among cultures is the solution to many problems. This vague statement can be interpreted in different ways, leading to different consequences. Nevertheless, if everyone does their best to achieve this goal, we will take a big step forward. Wherever different cultures, practices, or mindsets meet, everyone can take responsibility to maintain respect and an adequate discussion to find answers to the big tasks of our time. I want to follow one of my former lecturers at the German Jordanian University in Amman, who states that the right "understanding [of intercultural communication] provides opportunities to bridge gaps and embrace differences while empathizing with the wide range of differences present in our world" (Haddad 2021, p. 440). Intercultural communication is important in many aspects so that sustainable bridge building can be achieved on a global, national, local and individual level. This paper is the product of one of these bridge building events, the Autumn school "Balanced Sustainability in a Changing World", attended by participants from all around the globe. This environment of communication between cultures and generations is one positive aspect of an economic, technological and social globalization, which undoubtedly also has dark sides. Progressiveness in transportation, communication, and information exchange stands in contrast to an emerging exploitation of nature and people, political polarization, and western-centric knowledge creation. The last part is going to be important for this part of our text. Particularly with regard to the climate crisis and the enormous threat our world is facing, it is significant to gather as much knowledge as possible. Indigenous knowledge could have unimagined potential unrecognized by the broader society. We would like to present what we see as the new position of indigenous knowledge in the scientific discourse concerning the climate crisis and sustainable development goals. We want to emphasize, to further strengthen the role of indigenous knowledge in the scientific community, as well as the implementation in politics.

"Local and indigenous knowledge [also traditional knowledge] refers to the understandings, skills and philosophies developed by societies with long histories of interaction with their natural surroundings" (UNESCO w.d.). That is one of many definitions, in this case provided by the UNESCO's Local and Indigenous Knowledge Systems program (LINKS). We also recognize the discourse about terminology, but will mainly refer to the term "indigenous knowledge".

Next to UNESCO, a lot more NGOs and public institutions have already recognized indigenous knowledge as important for today's and future challenges. Also scientifically, you can see a positive process. *The Intergovernmental Panel on Climate Change* (IPCC), as well as the *Intergovernmental Science-Policy Platform on Biodiversity and Ecosystem Services (IPBES)* recognize the important role of indigenous knowledge in questions of the climate crisis. The *IPCC Special Report on the impacts of global warming of 1.5°C* (IPCC 2019) as well as the *Global Assessment Report on Biodiversity and Ecosystem Services* (IPBES 2019) implemented indigenous knowledge in their testimonies and called for joint actions with indigenous communities. Therefore, Indigenous knowledge is first and most importantly, scientifically recognized as a source of knowledge that could help us tackle the consequences of the climate crisis. Valuing this knowledge and recognizing it as crucial is the result of a scientific discourse over decades. And while the discussion continues in its complexity about specialized partial aspects, it is a concession. This concession to the indigenous community is tragic, as indigenous people are in particular affected by the consequences of the climate crisis. "Populations at disproportionately higher risk of adverse consequences with global warming of 1.5°C and beyond include [...] some indigenous peoples, and local communities dependent on agricultural or coastal livelihoods (high confidence)" (IPCC 2018, p. 9). It is important to take victims of the climate crisis seriously, and to value their knowledge regarding adaptation to the climate crisis consequences. This is not only important on a normative or moral level, but also scientifically comprehensible. "There is medium evidence and high agreement that indigenous knowledge is critical for adaptation, underpinning adaptive capacity through the diversity of indigenous agro-ecological and forest management systems, collective social memory, repository of accumulated experience and social networks" (IPCC 2018, p. 337). The scientific consensus thus seems to be clear in certain respects. Indigenous people are more affected by the consequences of the climate crisis. At the same time, indigenous communities have developed relevant adaptation processes. A more intensive examination could therefore be helpful. We would like to get practical in the following short part to make this relationship more vivid. We want to show that there are cases where clearly indigenous practices are combating the consequences of climate crises. We also want to counter critique, saying indigenous knowledge is locally bound and cannot be upscaled. We think the criticism is justified for some points, but not in general.

Today's consumer chains are a new-fangled concept. Modern agriculture is highly threatened by the consequences of climate crisis. A lack of resilience and monocultures make food sources susceptible. It may be helpful to adapt indigenous practices to create a more crisis-proof food production. "Throughout the centuries, indigenous peoples have developed agricultural techniques that are adapted to extreme environments, like the high altitudes of the Andes, the dry grasslands of Kenya or the extreme cold of northern Canada" (FAO 2017). This is a perfect example of how indigenous knowledge could help secure food supply and provide a valuable source to modify agriculture in the future.

But also, the food of indigenous communities itself could bring important benefits. Prof. Jan-Heiner Küpper presented the algae spirulina during the Autumn-School. It originated from Lake Chad, and is very rich in protein and has the ability to store a large amount of CO_2. UNESCO (2020) calls it "An example of nature-based solutions contributing to biodiversity conservation, poverty reduction and sustainable development in the Lake Chad area". This algae has not only the potential to tackle world hunger, it could also directly impact the climate crisis — an exciting research field and possible example for a hidden treasure of indigenous knowledge. However, when talking about the adoption of certain practices or even products from indigenous communities, we would like to strongly emphasize the responsibility mentioned in the introduction part.

Next to food security, indigenous knowledge could help enable integrative governance, integrate sustainable forests, conserve terrestrial landscapes, manage marine systems and so on. The IPBES Report (cl. 2018, XLVIII fff.) mentions many more fields where indigenous knowledge could be very beneficial, which we will not go into in detail anymore. We want to focus on a basic aspect, we can learn from indigenous communities. One of the most important lessons we can learn is to rethink our relation with our environment and to change our mindset. Indigenous communities have a fundamentally different understanding of their relationship to nature. A reconnection to mother earth on the basis of indigenous knowledge would be the development that could change everything. A social and global reorientation would again strongly correlate with the opening sentences on individual responsibility. This important change can also be driven by education and an adaptation of certain learning contents at the different levels. And there are already concepts trying to achieve this goal. Buen Vivir, "a Latin American concept based on indigenous ideas of communities living in harmony with nature, is aligned with peace; diversity; solidarity; rights to education, health, and safe food, water, and energy; and well-being and justice for all" (IPCC 2018, p. 480). This could be an interesting topic, people should get into more frequently.

2.1.3 *Ayahuasca: Learning and healing*

In this part of the work we want to focus on an elementary form of relationality with the nature: the shamanism. Shamanism has been regarded as one of the world's oldest religions, spiritual and healing practice (Pollock 2019, Winkelman 2002). Winkelman goes so far as to argue that this constitutes humanities' first theological and spiritual system. Unfortunately, due to the mission of colonization and indoctrination, this knowledge was relegated to beliefs, idolatries and myths, without scientific-theological validity (Boaventura Santos 2010).

However, shamanic practices have achieved a great modern resurgence. One of the most important and intriguing practices of shamanism is the induction of ASCs (Altered States of Consciousness).

These states of consciousness might be produced by many different techniques, as fasting, drumming, singing, chanting or the ingestion of psychoactive substances. The ASC happens through a slow-wave synchronization across brain systems, stimulating an integrative model of consciousness that enhances self-awareness and social identity (Winkelman 2000). It promotes a psychodynamics with the potential of healing and cognition process, involving self, attachment, social bonding, emotions and integrative brain functioning (Winkelman 2004).

Here we want to pay attention to a particular mode of knowledge and shamanic ASCs induction: the ritualisation, sacralisation and consumption of elements of the vegetable kingdom, such as plants and fungi for religious and medicinal purposes. Ethnobotanists believe that the use of psychoactive plants is as old as human beings (Furst 1992, McKenna 1993). Some authors consider this practice the origin of all religions (Gordon-Wasson 1992, Furst 1992, La Barre 1972). McKenna (1993), suggests that the ingestion of mushrooms containing indole-type alkaloids was one of the factors that helped to shape our brain as it is today.

The "sacred medicines" as they were often called, helped in a series of physical and psychological illnesses, syndromes linked to culture and community and social disharmony (Schultes and Winkelman 1996). Modern science has rediscovered the medicinal wisdom of our ancestors and, based on numerous researches, it has been shown that the use of plants with psychedelic properties can contribute in a revolutionary way to new methods and treatments in psychiatry.[2]

Some are the categories given to this type of substances, such as the terms hallucinogenic or psychedelic.[3] However, the religious dimension of the consumption of these substances has led Wasson, Hoffman and

---

[2] See: BOUSO, José; RIBA, Jordi. Ayahuasca and the Treatment of Drug Addiction. In: LABATE, Beatriz; CAVNAR, Nancy. The Therapeutic Use of Ayahausca. 2014, and PALHANO, Fernanda *et al*. The Therapeutic Potentials of Ayahuasca in the Treatment of Depression. In: LABATE, Beatriz; CAVNAR, Nancy (Org.). The 2014.

[3] Etymologically means manifestation of the mind. According to Osmond (1972), the experiences with these substances expand the limits of consciousness, broaden the vision and enrich the spirit.

Ruck, in El Camino a Eleusis (1980), to propose a very intriging term: "entheogen".[4] Entheogen is a neologism that means "God within us" or "full of God", that is, plants and substances that when ingested, would have the potentiality to manifest the divine within.

For the search parameter at this moment we want to stick to a specific substance: ayahuasca. It is a drink prepared from the cooking of a vine and the leaves of a shrub. Ayahuasca is a word of Quechua origin: Aya — soul, spirit and huasca — vine: vine of the soul.[5] It is an ancient[6] drink used by at least 70 indigenous peoples of the Amazon forest region. It is considered a sacred drink and is worshiped by the people who use it for its potentiality to promote religious ecstasies, its healing properties and access to the spiritual world.

A fundamental point is that ayahuasca is not understood as an object, but as a subject. Both plants used have a spirit that communicates with the person who ingests the drink. This may seem like an "epistemological heresy" (Albuquerque 2018), after all, how can a plant be subject and resource for knowledge?[7] The epistemic-cosmological debate that we propose here necessarily involves the rescue of a sensitive look at the spiritual dimension of nature, which sees all creation as revelation and endowed with divine substance.

The symbolism that ayahuasca carries is based on two main systems: learning and healing. It is therefore considered a teacher drink and a medicine drink. But what kind of knowledge is this? And what would be its way of healing?

---

[4] In Greek, "entheos" literally means "God within" or "full of God"; it is from the same root as "enthousiasmós", divine excitement. Combined with the root "gen", which denotes the action of becoming, this word composes the proposed term: "entheogen". It is a chemical substance, typically of plant origin, that is ingested to produce an unusual state of consciousness for religious or spiritual purposes (ENTHEOGEN 2007).

[5] This is the best known of the 42 names used to name the drink. "Caapi", "Yagé", "Kamarampi", "Honixua", "Hoasca" and "Daime" are some of the other names given to ayahuasca.

[6] From studies carried out by the Ethnological Museum of the Central University of Quito, Ecuador, it is estimated that the drink has been used for at least 2050 years (LUNA 1986).

[7] This is due to a process of epistemological turn, where for shamanic cosmologies, there is no conflict between nature and culture (BOAVENTURA 2010).

Knowledge would not be related to an intellectual or discursive type, but to a system of existential knowledge, a self-growth, self-awareness and a maturing of the soul. These are states of vision, revelation and illumination with meanings that allow an enrichment of the spirit towards the depths of truth, finding answers about the Ultimaty Mystery of life.

The disease is understood beyond "organic dysfunction, but is inserted in a framework of cosmic reference" (Monteiro da Silva 1985, p. 16). It would be above all, a sick soul that disconnected itself from the harmonious order and needs to be rescued. Healing takes place through a path of inner transformation, purification and re-signification of life.

Previous studies show that this quality of substances have the potential to enhance mindfulness-related capacities, such as decentering, non-judging, non-reacting and acceptance (Clavé *et al.* 2020). Other characteristics studied from the use of psilocybin show that it improves social approach behaviors and social interaction, social connectedness, emotional self-control and tolerance, prosocial attitudes/behaviors, healthy psychological functioning and positive personality changes (Arce and Winkelmann 2021). These experiences often show an enduring effect on individual and group well-being and society, even providing increased ecological and planetary care (Isham *et al.* 2022).

We point out that the self-experience and self-awareness through the entheogenic pathway that this sacred substances offers, simultaneously emancipates knowledge about oneself, but it goes beyond the individual territory and takes place in a cosmological dimension and in relation to the whole. A knowledge that is alive and dynamic, experiential and pragmatic and that deepens the spiritual-human-nature relationship, in a task of sacralization and re-enchantment of life on earth, providing the alternative ways for structuring new narratives and scientific perspectives, reinforcing that these ancient practices help us to improve our sense of care with nature and other beings.

2.1.4 *Implementing indigenous knowledge in urban regions*

If we want to establish a balanced sustainability in a changing, globalized and continuously growing world for us and for following generations, we have to keep our focus on the three major dimensions of sustainability: Economy, Environment and Social life. The concept of sustainability only works as long as all three dimensions are taken into account equally. In our changing world, it is almost impossible to accomplish a perfectly balanced system like this. Over the last centuries, our understanding of how we can live the best life, laid mainly on economic growth. The environment and society have suffered from this worldview and have been exploited to support the economic sector.

Already in the late 20th century, with the publication of the "Limits to Growth" of the Club of Rome, it became clear that global sustainability, which also sets limits to economic growth, is necessary. Nevertheless, the path of maximum economic growth has been pursued to this day, while climate protection and social justice have been regarded as secondary. Although these two dimensions are increasingly taking up space in political and public discussions, goals such as the SDGs or the decisions of the climate conferences in Paris (year) are only tentatively being implemented. In a changing world — socially, economically and climatically/environmentally — a drastic change in our way of thinking and our principles is needed.

The consequences of our actions are becoming increasingly apparent due to climate change, loss of biodiversity and growing social injustices. Particularly affected by the environmental and climatic changes are the population groups that live in close ties with nature and have contributed the least to these global challenges — the indigenous communities.

Whilst in the eyes of western culture the physical landscape is mainly an ensemble of natural forms and natural phenomena, the meaning of landscape for indigenous people is deeper, more complex and much more spiritual. The surrounding landscape or country, from an indigenous view, can be defined as:

Country [...] is multifaceted, it includes the physical, non-physical, linguistic, spiritual and emotional. It includes self, and feels emotion as we do. If we are sick culturally our Country becomes sick, so maintaining culture maintains Country. [...] Country is family, incorporating its animals, plants, landforms and features right down to the smallest of things like a grain of sand. (Nicholson and Jones 2018, p. 379)

This understanding of nature and one's own environment, the spirituality and connectedness is missing in the mindset of non-indigenous people. Indigenous knowledge has a transformative power that can help us with a variety of cultural contexts, and especially with the human consciousness (Semali and Kincheloe 2011). Indigenous knowledge is gained almost entirely through oral practices such as stories and legends, but also through dances, symbols and rituals. The knowledge is based on centuries of experience, exploring, refinement of techniques, listening and feeling and it gets passed on from generation to generation (Semali and Kincheloe 2011). In order to have knowledge shared by indigenous peoples, we must learn to listen to them and abandon our very objective methods of Western science and instead allow diversity and other beliefs (Simonds and Christopher 2013).

Since we have a global trend of urbanization with many urban regions growing uncontrollably fast into Megacities with over 10 Million inhabitants, the urge to find the right ways of planning cities is getting more and more urgent. We believe that indigenous knowledge and the indigenous spirituality can help us make cities resilient and livable, and not only strengthen awareness of the natural environment, but also improve our communities and social structures.

Cities around the world face a large number and variety of challenges. Globally, cities have been subject to steady growth for several decades. By 2050, up to two-thirds of the world's total population will live in urban regions (Hunter *et al.* 2019). The population growth therefore also means an expansion of the city in width, height and density. The expansion results in large-scale sealing, which in turn leads to urban challenges: loss of biodiversity and ecological services, impairment of the natural water cycle, or intensification of urban floods, heat islands and heat waves (Pearlmutter *et al.* 2021). In addition, there is air and noise pollution due to high traffic volumes, high energy, drinking water and food requirements, as well as health impairments. Social exclusion and injustices are also exacerbated by urban growth (Keivani 2009).

Our way of life, not only in cities but generally in our society, is designed for continuous economic and also personal growth. The rapid growth of cities also supports the effects of acceleration and alienation in our lives. The rural happiness paradox, a phenomenon revealed worldwide, shows that the population of rural areas is happier and reports a higher level of well-being in their living environment than inhabitants of urban areas. Nevertheless, more and more people are moving from the countryside to the city. The exact reasons for this paradox are unclear (Sørensen 2021). However, the feeling of being part of a community and being in close contact with people can help residents feel more comfortable in their surroundings. Since the sense of community is significantly higher in rural areas than in highly anonymous cities, this could be a possible explanation for the paradox.

Indigenous people maintain a special cohesion within a tribe. This sense of community, the support and close connection to fellow human beings, should be more strongly integrated into social life in cities.

But how can we learn such behavior? A project that is already taking a step in the direction of this social thinking has been started by WOGENO in Munich, Germany. The organization bought a vacant residential building and renovated the building with the help of the future residents. Each of the residents had to spend a specific amount of hours on renovation duties and received an affordable apartment in return. Through the cooperation of all residents, a strong cohesion and a strong sense of community developed — a big step towards a socially sustainable future (WOGENO n.d.).

Another example that can permanently change the mentality of citizens in the future is the *Superillas* in Barcelona, Spain. These Superblocks are areas in the city where the traffic was calmed down and the focus of mobility has been set on biking, public transportation or walking. There was a lot of critique at the beginning of planning these blocks, and there still is but a lot of people also cherish the more quiet, safer and greener living situation. Especially children have the chance to use the now free public spaces for play and personal development. In large cities such as Barcelona, which are struggling with very high levels of car traffic and offer little safety on the road, this is a positive development for the social life of all generations. Another positive effect on the rising generations is that they are shown from an early age that a life without their own car is possible, unlike generations who grew up with private cars as a means of transport and now have to actively deal with a change in mentality towards mobility in cities. This change in basic mentality can lead to a more sustainable lifestyle for the rising generations.

These blocks can not only help us achieve a change in social mindset, but also build a deeper connection with the nature around us. Within the blocks, but also in other cities on house roofs, balconies or free-standing areas, urban gardening can find a place. Indigenous people possess rich knowledge of native nature, plants and herbs. Although the knowledge is very specific to a region and the climatic conditions and can therefore only be applied in cities with similar local conditions, urban gardening can bring a variety of opportunities to cities. It can provide general understanding of what vegetables, fruits or grains are in season, what can grow locally and reduce mass consumption and production (Settee 2008). Further urban gardening can promote health, physical activity, local climate and reduce long-distance transportation of food (Guitard *et al.* 2012). To maintain the mentality of indigenous people we have to keep in mind that urban gardening should be primarily a contribution to improving the natural environment, secondly for strengthening the community and lastly it should be seen as something where economic profit can be made of. In the last part of this text, the opportunities of urban gardening are examined more closely.

A great economical approach towards a more sustainable future was made by Amsterdam during the COVID-19 Pandemic. The city wanted to transform its entire economic system and follow a fully circular economy by 2050. The underlying economic concept is based on the Doughnut Economy by Kate Raworth (2017). In this concept, two rings, one external and one internal, limit the political and economic scope for action. In the middle is a hole — as in the pictorial symbol of the doughnut. The inner circle represents all basic social and human needs: access to water, health care, security, etc. Since a certain limit must not be undercut in these basic needs, the model has a hole in the middle. Outwardly, the model is limited by the planetary limits — climate change, biodiversity loss, substance inputs, etc. The model shows that infinite economic growth is not possible, and that attention must be paid to the natural limits and signals of nature in order to secure human livelihoods (Raworth 2017). Ailton Krenak, a member of the Krenak tribe in Brazil and indigenous activist, stated, after being asked about the effects of the global COVID-19 Pandemic:

> In the first moment there were pandemics, people harboured a certain hope that humanity would come out of it better, and that's not what actually happened. I myself believed back in the first year that at that stop of everything, everyone would have a moment of reflection and redirection to very obvious things. For example, the idea that the economy is not the only indicator of life success for a population. If capitalism put a brake on this dynamic of the global financial system then environmental issues would gain relevance. (Krenak n.d.)

Even if Krenak's assessment applies to most parts of the world and hardly any change of heart has taken place among the general population, the implementation of this economic principle, should it actually prevail in Amsterdam, would be a big step towards his ideas and hopes. It would be a great success for the design of a sustainable future and an approach to the understanding and relationship between humans and nature in the indigenous sense.

One further important lesson we can learn from indigenous tribes is how we deal with natural events and disaster risk reduction. Indigenous tribes are at high risk to be affected by natural hazards. With climate change and intensifying natural events the risk for these people is rising. But over thousands of years they were able to cope with these hazards because of their strong connection and understanding of the nature (Ali *et al.* 2021). In cities we mainly rely on technical infrastructure. The rapid growth of cities pressures citizens to move into naturally endangered regions but through technical infrastructure and intrusions in natural systems, we protect ourselves and we feel safe. But with bigger natural events happening, the technical infrastructure can break and the damage will be even more severe. It is important to revive natural safety systems and buffers like wetlands, floodplain forests or coastal areas and have technical infrastructure as support but not as the main safety system. The bigger and the more diverse the natural buffer is, the more resilient it will be to different kinds of hazards: floods, droughts or wildfire. We have to start implementing and working together with ecological systems instead of trying to work against nature. This is the mindset we should adopt from indigenous people.

Of course, we cannot live in the globalized megacities of the world in the same way as indigenous people do, but in the future we can set our focus more on the environment and strong social connections and communities instead of promoting solely economic growth. With being more mindful, feeling more and allowing more spirituality in our world, we may reach a path to a sustainable global future.

2.1.5 *Inside urban gardening*

The city lifestyle has focused on how residents can use the facilities of a city with maximum efficiency of productivity. Consequently, it is hard to find green farms and gardens except in public parks. At the same time, there are often small pieces of land where trees or shrubs are planted. To build a sustainable future, we can set our focus more on the environment instead of only economic growth. Indigenous knowledge can be used to guide urban gardening which is the practice of cultivating and distributing plants in urban areas. It can reflect varying levels of economic and social development. And it has some benefits both for the public and the individuals. For example, urban gardening can boost biodiversity, and build the connections of the individuals and the nature. Repurposing unused city land for public garden, and creating more community gardens specifically near the high-density housing can protect the environment.

Indigenous people often have a strong feeling of being part of a community and be in close contact with other people, which is hard to attain in the city. Urban gardening can act as a way to raise the sense of community in highly anonymous cities. We could expect that public urban gardening connects the residents of a municipality so that many anonymous neighbors get in touch and recover their local community. It could prevent dehumanization in the local community, which is largely relying on anonymity. Residents joining public urban gardening have a certain level of freedom to organize the types of plants they plant. The more urban gardeners, the better biodiversity. Further, it lets the residents actively take part in city planning of where they are living. Not only for the people gardening the land but also residents passing by they could enjoy watching a diverse range of gardens and farms, not just the planned monotonic shrubs and trees. In short, the connection between individuals and between nature and individuals could be recovered in everyday life.

However, the practical implementation of the construction of the public urban gardening can be challenging. Citizens who are willing to join gardening are enthusiastic about environment protection and with much interest. Some of them may have the experience of gardening, while others would have never grown plants in the city. Also, they may lack some professional skills with regard to how to raise the plants. We could use indigenous knowledge to help settle these problems. For example, what vegetables or grains are in season, build circular agriculture systems to reduce consumption and pollution, how to apply moderate fertilizer. By holding regular meetings, the local community can communicate and share the experience of gardening and also strengthen their connections. The community development includes bringing people together, building connections, and bridging available resources with the people who need them. People who are either looking to learn how to garden or looking for help when seeking gardening resources can join the community.

We noticed that there is a "generation gap" that now exists when it comes to the knowledge of how to garden, and some young people do not know much about indigenous knowledge. So, it is necessary not only to provide the information on how to grow plants, but also to create a cultural atmosphere to accept the idea of indigenous knowledge. Incorporating gardening into the school system was also an actionable idea. It is not only an educational endeavor, but also a "cultural thing". Having hands on gardening experience would be an engaging environment to get close to nature and may benefit them for years to come. Lastly, the university could also be involved through horticultural or teaching internships (Martin and Wagner, 2018).

Here is an example of public urban gardening. Singapore is known as a city in a garden. Its transformation is related to the government policy. Singapore's former prime minister, Mr. Lee Kuan Yew, had a vision of making Singapore into a Garden City in 1967. This required government and private agencies to reserve spaces for trees and vegetation in their projects and buildings. Citizens' environmental and ecological education was improved. As a result, the city has seen its green areas continue to grow. A large network of tree-covered and pedestrian corridors connect parks to one another. Now it is a city with abundance of lush greenery and a clean environment to make life more sustainable for the country and the nation.

As these green spaces grew, so did the population of Singapore. This posed a challenge, since the city-state has a high population density. The solution to continue creating green spaces despite the increased population was to combine architecture and vegetation, to build eco-friendly building and vertical gardens. It is normal to find plants on the top and the sides of buildings (like cascading gardens) and also inside the buildings. One of the finest examples of the union between architecture and nature on the island is the Jewel Changi airport, which combines natural light, water and green spaces. One of Singapore's main strengths in following this line of action in the future is the environmental awareness of its citizens under the guidance of the indigenous knowledge.

2.1.6 *Conclusion*

Indigenous knowledge was built up over centuries. It should undoubtedly be respected as a valuable source of knowledge with potential improvements and even life-saving measures for an individual as well as on the global level. Scientifically it was recently acknowledged as a possible tool to adapt to climate crisis consequences. Local and national cooperations with indigenous people could reduce the consequences of this crisis. We therefore see a positive trend in acknowledging and respecting indigenous knowledge scientifically. Adapting indigenous knowledge and rethinking our connection to our environment could have a great impact on individuals as well as revolutionize the way we think about urban life. A respectful cooperation, locally and globally, as well as further scientifically with social acknowledgement, could lead to an exchange of knowledge and maybe to the discovery of internationally relevant practices. We are well aware that many indigenous practices cannot be applied globally. In our view, that should not be the ultimate goal anyway. Nevertheless, rethinking one's relationship with nature and the environment is something that anyone can do. We believe the discourse in total as well as individual cases provide an interesting field of research for different disciplines. This is also underpinned by our three different approaches in this text. We see the potential is not exhausted at all. Especially in the discussion of the autumn school, we realized that we are not only running out of time regarding the climate crisis, but also, regarding indigenous knowledge, that is being lost every day. It shows the actuality and relevance of this topic. But we are also convinced that indigenous communities who have experienced marginalization and stigmatization owe nothing to the international community, maybe not even to the majority society of the country they live in. It is therefore important, and we come back to the topic of responsibility, to respectfully ask for the help of indigenous communities, to offer cooperation on an equal footing and to act responsibly. Especially an interdisciplinary approach of climate scientists, indigenous knowledge holders and social scientists, concerned with intercultural communication could be fruitful. It is important to establish a holistic concept, acting together, collecting as much knowledge as possible and finding the best solutions.

References

Albuquerque, M. B. B. Pedagogia da Ayahuasca: Por uma decolonização epistêmica do saber. *Arquivos Analíticos de Políticas Educativas* 26, no. 85 (2018): http://dx.doi.org/10.14507/epaa.26.351.

Ali, T., Buergelt, P. T., Paton, D., Smith, J. A., Maypilama, E. L., Yungirrna, D., Dhamarrandji, S. and Gundjarranbuy, R. Facilitating sustainable disaster risk reduction in indigenous communities: Reviving indigenous worldviews, knowledge and practices through two-way partnering. *Int. J. Environ. Res. Public Health* 18, no. 3 (2021): 855. https://doi.org/10.3390/ijerph18030855.

Arce, J. M. R. and Winkelmann, M. J. Psychedelics, Sociality, and Human Evolution. *Front. Psychol.* 12 (2021): 729425.

Bouso, J. C. and Riba, J. Ayahuasca and the Treatment of Drug Addiction. *In*: Labate, B., Cavnar, N.: The Therapeutic Use of Ayahausca. Berlin/Heidelberg: Springer-Verlag (2014).

Clavé, E., Soler, J., Elices, M., Franquesa, A., Álvarez, E. and Pascual, J. C. Ayahuasca may help to improve self-compassion and self-criticism capacities, *Hum Psychopharmacol Clin* 37, no. 1 (2021): DOI: https://doi.org/10.1002/hup.2807.

FAO. 6 ways indigenous peoples are helping the world achieve #ZeroHunger. (2017): Available: https://www.fao.org/zhc/detail-events/en/c/1028010/.

Furst, P. Hallucinogens and Culture. Chandle & Sharp, United States (1992).

Gordon-Wasson, R., Hofmann, A. and Ruck, C. El caminho a Eleusis: una solución al enigma de los mistérios. Ciudad del México: Fondo de Cultura Economica (1980).

Guitart, D. P. and Byrne, J. Past results and future directions in urban community gardens research. *Urban Forestry & Urban Greening* 11 (2012): 364–373. https://doi.org/10.1016/j.ufug.2012.06.007.

Haddad, E. The Importance Of The Study Of Intercultural Communication As A Social Science. *Dirasat, Human and Social Sciences* 48, no. 1 (2021): 431–441.

Hunter, R. F., Cleland, C., Cleary, A., Droomers, M., Wheeler, B. W., Sinnett, D., Nieuwenhujisen, M. J. and Braubach, M. Environmental, health, wellbeing, social and equity effects of urban green space interventions: A meta-narrative evidence synthesis. *Environmental International* 130 (2019).

IPBES. Global assessment report of the Intergovernmental Science-Policy Platform on Biodiversity and Ecosystem Services. [Brondízio, E. S., Settele, J., Díaz, S., Ngo, H. T. (eds)]. IPBES secretariat, Bonn (2019).

IPCC. Global Warming of 1.5°C. An IPCC Special Report on the impacts of global warming of 1.5°C above pre-industrial levels and related global greenhouse gas emission pathways, in the context of strengthening the global response to the threat of climate change, sustainable development, and efforts to eradicate poverty [Masson-Delmotte, V., P. Zhai, H.-O. Pörtner, D. Roberts, J. Skea, P.R. Shukla, A. Pirani, W. Moufouma-Okia, C. Péan, R. Pidcock, S. Connors, J.B.R. Matthews, Y. Chen, X. Zhou, M.I. Gomis, E. Lonnoy, T. Maycock, M. Tignor, and T. Waterfield (eds.)]. Cambridge University Press, Cambridge, UK and New York, NY, USA, (2018): 616 pp.

Isham, A., Elf, P. and Jackson, T. Self-transcendent experiences as promoters of ecological wellbeing? Exploration of the evidence and hypotheses to be tested. *Frontier in Psychology* 13 (2022):105147.

Keivani, R. A review of the main challenges to urban sustainability. *International Journal of Urban Sustainable Development* 1, nos. 1–2 (2009): 5–16. https://doi.org/10.1080/19463131003704213.

Krenak, A. (n.d.). The Future is Ancestral. *In:* Campos, C. (ed.): *Where the leaves fall, 8.* Available: https://wheretheleavesfall.com/explore/article-index/the-future-is-ancestral/.

Mabit, J. Produção visionária da ayahuasca no contexto curanderil da Alta Amazônia Peruana. *In*: Labate, B. C., Sena Araùjo, W.: O Uso Ritual da Ayahuasca. Campinas: Mercado de Letras (2002).

Martin, W. and Wagner, L. How to grow a city: Cultivating an urban agriculture action plan through concept mapping. *Agriculture and Food Security* 7, no. 1 (2018): Scopus. https://doi.org/10.1186/s40066-018-0186-0.

Mckenna, T. Food of the gods - the search for the original tree of knowledge. Bantam Books (1992).

Monteiro da Silva, C. O Palácio de Juramidam – Santo Daime: um ritual de transcendência e despoluição. Dissertação de mestrado em Antropologia Cultural. Recife: UFPE (1983).

Nicholson, M. and Jondes, D. Urban Aboriginal Identity: 'I Can't see the Durt (Stars) in the City'. *Remaking Cities: Proceedings of the 14th Urban History Planning History Conference* (2018): 378–387.

Osmond, H. Sobre alguns efeitos clínicos. *In*: Bailly, J.-C., Guimar, J.-P. (Orgs.) Mandala: A Experiência alucinógena. Rio de Janeiro. Civilização Brasileira (1972): 42–69.

Palhano Fontes, F., Alchieri, J. C., Oliveira, J. P., Soares, B. L., Hallak, J. E. C., Galvao-Coelho, N. and de Araujo, D. B. The Therapeutic Potentials of Ayahuasca in the Treatment of Depression. *In*: Labate, B., Cavnar, N. (Org.). The Therapeutic use of Ayahausca. Berlin/Heidelberg: Springer-Verlag (2014): 23–39.

Pearlmutter, D., Pucher, B., Calheiros, C., Hoffmann, K., Aicher, A., Pinho, P., Stracqualursi, A., Korolova, A., Pobric, A., Galvao, A., Tokuc, A., Bas, B., Theochari, D., Milosevic, D., Giancola, E., Bertino, G., Castellar, J., Flaszynska, J., Onur, M., Mateo, M., Andreucci, M., Milousi, M., Fonseca, M., di Leonardo, S., Gezik, V., Pitha, U. and Nehls, T. Closing Water Cycle in the Built Environment through Nature-Based Solutions: The Contribution of Vertical Greening Systems and Green Roofs. *Water* 13 (2021): 2165–2198.

Pollock, D. Shamanism. *In*: OXFORD Bibliografies. (2019): Available: https://www.oxfordbibliographies.com/display/document/obo-9780199766567/obo-9780199766567-0132.xml.

Porter, L., Hurst, J. and Grandinetti, T. The politics of greening unceded lands in the settler city. *Australian Geographer* 51 (2020): 221–238. https://doi.org/10.1080/00049182.2020.1740388.

Purvis, B., Mao, Y. and Robinson, D. Three pillars of sustainability: in search of conceptual origins. *Sustainability Science* 14 (2019): 681–695.

Raworth, K. *Doughnut-Economics. 7 Ways to Think Like a 21ˢᵗ Century Economist.* Chelsea Green Publishing (2017).

Santos, B. de S. and Meneses, M. P. (Orgs.). Epistemologias do Sul. São Paulo; Editora Cortez (2010): 637.

Schultes, R. and Winkelman, M. The principal American hallucinogenic plants and their bioactive and therapeutic properties. *In*: WINKELMAN, M. (Org.). *Yearbook of Crosscultural Medicine and Psychotherapy*. Berlin (1996): 205–240.

Settee, P. Indigenous Knowledge as the Basis for our Future. *In:* NELSON, M., K. (ed.) *Original Instructions: Indigenous Teachings for a Stustainable Future*. Smithsonian Libraries (2008).

Simonds, V. W. and Christopher, S. Adapting Western Research Methods to Indigenous Ways of Knowing. *Am J Public Health* 103 (2013): 2185–2192.

Sørensen, J. F. L. The rural happiness paradox in developed countries. *Social Science Research* 98 (2021): https://doi.org/10.1016/j.ssresearch.2021.102581.

Timothy, L. and Walter, C. *Religious Implications of Consciousness-Expanding Drugs. Relig. Educ.* Vol. LVIII, no. 3 (1963): 252.

Troeltsch, E. The Social Teaching of the Christian Churches. 2 vols. Nova York: Harper e Row (1960).

Weber, M. A psicologia social das religiões mundiais. *In*: Ensaios de sociologia. 5 ed. Rio de Janeiro: Zahar (1982): 309–346.

UNESCO. Spirulina, a miracle ingredient in Lake Chad. (2020): Available: https://www.unesco.org/en/articles/spirulina-miracle-ingredient-lake-chad.

UNESCO (w.d). Local and Indigenous Knowledge Systems (LINKS). Available: https://en.unesco.org/links.

Winkelman, M. Shamanism The neural ecology of consciousness and healing. Westport, CT: Bergin and Garvey (2000).

Winkelman, M. Shamanism as Neurotheology and Evolutionary Psychology. *American Behavioral Scientist* 45(12) (2002): 1873–1885.

Winkelman, M. Shamanism as the original neurotheology. *Zygon* 39(1) (2004): 193–217. https://doi.org/10.1111/j.1467-9744.2004.00566.

WOGENO (n.d.). *Metzstraße 31*. Available: https://www.wogeno.de/haeuser/haeuser-im-bestand/metzstrasse.html.

3 The eco-matrioska — *Social transition from a multilevel and integrated perspective*

Stella Drebber
Leuphana, Universty of Lüneburg, Germany

Yunfei Fan
School of Psychological and Cognitive Sciences, Peking University, China

Matteo Sesia
Department of Psychology, University of Torino, Italy

Darja Podvigina
University of St. Petersburg, Russia

3.1 Abstract

The climate crisis already is and will continue to be one of the toughest and most urgent challenges humankind must go through. The problems that come with it are becoming more and more pressing, nonetheless many people are still denying it, and much more is going on with their lives like nothing was. How is that possible? How to change our approach to the issue? The present paper tries to give answers to these two questions using a multilevel and integrated method, approaching the problem from different perspectives. Firstly, the argument will be dealt with from a cultural point of view: it will be shown how the culture and the societal norms influence and shapes personal behaviors and decision-making processes. Secondly, the topic will be addressed using a social psychology perspective: in this case, the focus will be stressed on how groups and group norms influence individuals and their decisions. In the end, the perspective will be narrowed down to individuals and how they individually behave, including some of the biological mechanisms of behaviour. It will be stated that people tend to avoid thinking about climate change, by psychological distancing themselves to avoid the anxiety that comes from it. Since acting greener is the more difficult option, people stick to their behaviors, condoning to the cognitive dissonance theory, instead of them reshaping their beliefs accordingly. This paper will focus on ways to change this.

3.1.1 *Introduction*

Climate change poses growing problems for humankind and requires responses at a wide range of levels, from individual to fundamental social changes. One variable that can drive sustainable development is the choice of sustainable options. The following text focuses on the question, how and why individuals make decisions, taking three levels of analysis, the cultural influences, in-group influences on concepts and behavior and the individual decision-making process (and their integration), into account. We will also look at the mechanisms that underlie cooperation, which is a necessary basis for sustainable behaviour. While keeping a psychological perspective and including factors such as external influence or anxieties around decisions, the question, how more sustainable choices (especially regarding bike usage) can be achieved, without exerting more pressure on individuals, is to be approached. The aim of the present paper will be theoretical more than practical, but it has been written in thematic exchange with our colleagues who contributed the following paper [Bicycle as a main transportation mode], therefore the two can be considered synergistically.

3.1.2 *From the cultural influence on decision making to the bike use within a mobility culture*

Making decisions such as riding the bike to work or using the car is individual, however cultural, and societal aspects have an influence on decision making processes and the available choice-set. All cultural differences depend on context, however general differences can be observed, regarding the influence of social contexts on decisions.

Individualistic cultures view each person as an independent entity, with great autonomy and free agency to pursue personal goals. Uniqueness and self-expression are generally valued and the accommodation of others or collective goals not the main focus (Yates 2016). Collectivistic cultures however view the "self" as a part of a whole. Personal behavior often gets adapted depending on the social context and to pursue group harmony. The focus lies on the groups' goals and their achievement. People in collectivistic cultures often weigh the input from others more (Yates 2016). Additionally, some cultures are tighter, with socially more strictly enforced norms, while others have fewer norms or view the violation of those less problematic, which identifies them as more loose (Yates 2016). When going a step deeper and focusing on the different think-processes, two types can also be identified. Holistic thinking, on the one hand, is characterized by attention to context and the conviction that everything is in a constant flux. Analytic thinking on the other hand, starts from the basic assumption, that the world is stable and predictable, with a focus on the main object (Yates 2016). When the decision-making-processes of individuals or groups are analyzed or ought to be influenced, this background knowledge has to be considered in order to better understand reasonings, because "Evidence from neuroscience also supports the constancy of cultural mindsets to some degree; repeated ways of thinking leave physical effects on the brain (Kitayama and Uskul 2011)" (Yates p. 108).

This cultural mindset changes how people interpret the same information, however, the choice-set or the presumed choice-set also plays an important role in this interpretation: "A poor choice set does not allow the decider to even contemplate the best possible option because it is not contained in the set for consideration" (Yates p. 110).

Two other characteristics, influenced by cultural differences are firstly: overconfidence, which stands in the way of questioning and changing one's own positioning and first impression. Secondly, one's feeling about losses, which influences risk aversion or risk affinity and therefore the decision based on the calculation of the risk (Yates 2016). In order to weigh the risks and possible solutions, they must be clearly identified, which can be already problematic coming from psychological research, which shows that individuals tend to discount uncertain future events when making trade-offs between cost and benefits. This is especially problematic regarding the climate change discourse. Most discourses on this topic heavily focus on losses, be it the loss of polar bears, the loss of biodiversity as a risk of climate change or the association of climate change solutions with immediate losses for society e.g. reduced car usage.

Going beyond these psychological aspects, members of one cultural community share understandings in areas such as communications, relationships, history, and language. "These shared understandings derive from historical, ecological, political, economic and media forces as well as negotiated or constructed understandings among residents and other constituencies" (Schensul 2009). Change in those systems can be brought upon on the macro-, exo-, meso- or microlevel via different ways, be it confrontation, interaction or through change agents, however a focus on the shared understandings, for example, regarding common media outlets should be considered (Schensul 2009). However, "multi-level interventions introduced in one community are likely to 'look' different in other communities, where policies, community alliances and population composition are different" (Schensul 2009). Continuing this thought, innovation and change are often best introduced in a local cultural form and from there extended to a national movement, which then introduces other local effects (Schensul 2009).

When bringing these theoretical findings on culture and decision-making together with recent questions around the environment, it becomes clear that many sustainability issues connect with these decision-making processes. Eco-anxiety and eco-distress, defined as "the chronic fear of environmental cataclysm that comes from observing the seemingly irrevocable impact of climate change and the associated concern for one's future and that of next generations" (Iberdrola 2022) by the American Psychology Association (APA), for example, influence a person's view on available choices as well as possible losses. Although we are not able to diagnose eco-anxiety yet; the recognition of it and its complex psychological effects are increasing and having an exponentially growing impact on children and young people (Gregory 2021), it can often influence the view on togetherness and societal responsibilities, letting individuals take the blame for the environmental situation on themselves (Bischoff 2022).

One discourse in the field of solutions to climate change revolves around bicycle usage as an eco-friendly alternative to many other transportation forms. While physical determinants are one of the basic requirements for the use of bicycles, cultural factors also play a role (Pelzer 2010). In countries such as the Netherlands, which are known for an extensive bicycle-culture, the bike has become a natural and wide-spread transportation mode, "because it is part of the upbringing in a lot of Dutch households and embedded in institutions and standards" (Pelzer 2010). This shared identity becomes visible when compared to "the other", in this context immigrants, in the Netherlands (Pelzer 2010). The embedding of bicycling as a cultural phenomenon is influenced by the norms and values surrounding it as well as practical conditions such as good infrastructure, laws in favor of bicyclists or stakeholders who advocate bicycle use (Pelzer 2010). Today, it can be regarded as a symbol for national identification.

The findings show that "to come to a satisfactory understanding of 'bicycle culture' it is necessary to explore both the material and socially constructed properties of bicycling. The dimensions of the physical environment and the socially constructed dimension (mobility culture) are far from mutually exclusive and interact in a complex way. [...] There's no 'one-size-fits-all' cycling stimulation policy. It is pivotal to be sensitive to the cultural context of a city" (Pelzer 2010). Using the given groundwork, in the following, different options and mechanisms are going to be analyzed to approach the questions around a sustainable change and the decisions for sustainable options, first and foremost for using the bike from a social psychology viewpoint.

Zooming in from the cultural to the social psychology aspect, we observe that individuals are shaped in the group they belong to, both on the concept and behavior level. Individuals live among certain groups, which provide information to them as well as restricting the content they could reach. Especially when the recommender system has been adopted by most of the mainstream social media, people are more likely to live in an information cocoon designed for them. Therefore, the information one could reach and the concept deriving from that are shaped by the social environment. However, even when sharing similar ideas, people would still make different choices depending on different circumstances. Two factors, which influence decision-making are the objective conditions and the choices made by others. These factors are considered, while this part is focused on the social psychology theories which provide us some insight on sustainable actions, such as responsibility diffusion and information cocoons. Combined with the Green Nudge theory the analysis revolves around the question, how social psychology theory could help individuals to reach more sustainable information and lead a greener life.

3.1.3 *Information cocoons and recommender algorithms*

Information cocoons is a concept advanced by Sunstein (2006) referring to the phenomenon that people tend to see information which are in line with their own opinions through the interplay of their own preferences and recommender systems. Therefore, their perspectives are narrowed down just like they are chambered in information cocoons which keep the conflict content out (Sunstein 2006). This effect not only narrows down our horizon but makes us more likely to follow those who share the same idea, when making decisions, despite our real abilities. This can be seen in different experiments: some researchers let their participants evaluate financial assets, while giving credit to other participants according to their performance. Intuitively, participants should give credits to others, after they receive the feedback and find their evaluation was precise. However, the judgement of others' expertise was being made before they received the feedback. Participants took their own decision into consideration when evaluating others' ability. They would give those who shared similar evaluation on the financial assets much more credit than those who had conflicting ideas, even though they made accurate evaluations (Boorman *et al.* 2013). Moreover, the accreditation triggered by similarity could even be generalized to different fields. Researchers recruited participants to do the task of shaping discrimination while giving them the opportunity to ask for help from other participants, who they had exchanged political opinions with. The results suggested that participants tended to ask those for help, who shared their political opinion. They were also influenced more by them, even though they knew that they were not that good at shape discrimination tasks (Marks *et al.* 2019). These psychology theories and studies suggest that individuals are influenced easier by like-minded others and that their opinions are, to some extent, restricted by their own preference.

Based on these phenomena, researchers developed fMRI studies to check the fundamental mechanism underlying this type of in-group influence. Participants were divided into three groups by a minimal group paradigm approach (Tajfel *et al.* 1971), and instructed to conduct a dot pattern perceiving task with a MRI machine. They were asked to report the number of dots giving the artificial information of the results from their group members or people from other groups. The results showed that participants were more easily persuaded by in-group results and found out, that in line with other research, the activity in the caudate was specifically triggered by in-group conformity (Zaki *et al.* 2011, Stallen *et al.* 2013). The posterior superior temporal sulcus (pSTS) was also proved to be exclusively included in in-group influence (Stallen *et al.* 2013). Besides, the pSTS is always related to the cognitive capacity of mentalization, so these studies also suggested that individuals are more likely to take the perspective of their in-group members than out-group members (Freeman *et al.* 2010, Heatherton 2011).

Recently, with the hit of the internet and social media, the in-group influence could be exaggerated by the recommender algorithms. Due to the high profits made by recommender systems, they have been applied to almost every corner of the internet, let alone social media platforms. Basically, a recommender system is an information filter, selecting preferred information for each customer. Using machine learning or other approaches with artificial intelligence, the recommender algorithms could take factors like the search history, the users' profile, and the preferred content into account to form a hybrid recommendation approach to personalize recommendations (Hoekstra 2010, Gomez *et al.* 2015). Based on the knowledge of the tendency in decision-making, some algorithms even mimic the spiral of silence (Matthes 2015) to filter information more effectively (Lin *et al.* 2022). Based on these facts, social media is a lens exaggerating the in-group influence, and thus could be used as a powerful campaign tool to promote sustainable options.

3.1.3.1   Responsibility diffusion

In addition to sufficient information and the attitude of others, the responsibility we should take for the decision we have made is also a crucial factor we would consider. The condition is a little bit complex, since sometimes we hesitate to take the responsibility, but sometimes we strive for something because we feel responsible. Especially, when we act in groups, responsibility diffusion might have a considerable influence on our behavior. People tend to take less responsibility when they have co-actors which is called the diffusion of responsibility. In other cases, when there is an authority to provide a default option or give some command, people also have the tendency to pass the buck to the authority and reduce their own responsibility (Hayashida *et al.* 2021). The study of Milgram (1963) showed that participants obey the instruction to harm the "learner" because their responsibility was diffused from themselves to the authoritative experimenter.

In other words, responsibility could either diffuse to in-group members or the authority, who could, to some extent, lead people to a different behavior, compared to when they make their decision alone. The most common effect of responsibility diffusion is to lead individuals to act in a more aggressive way and take less evaluation on the results (Falk and Szech 2017, Li Peng *et al.* 2010). However, diffusion of responsibility could also reduce the pressure of making a decision (El Zein *et al.* 2019).

Besides, subjects would pass the buck to others when the cause of the bad results is ambiguous (Mezulis *et al.* 2004), and clarifying the causality is not enough to prevent them from doing so (Hayashida *et al.* 2021). Nevertheless, when the frame of buyer/seller interactions were applied to the experiments, there was no significant difference, when people made socially responsible decisions individually or in a group (Irlenbusch and Saxler 2019). So, introducing the mindset of marketing might reduce the diffusion of responsibility and help people evaluate cost and benefit more carefully.

3.1.3.2 The connection between social psychology and nudge theory

Combining the aforementioned knowledge from social psychology and many other fields, nudge theory becomes relevant. It is a concept that refers to changing the environment slightly, to induce people to make a certain choice automatically (Tagliabue and Simon 2018). For example, one typical approach of green nudging is to offer the desired option as the default one. Individuals would automatically receive the default option without any effort, which to some extent could help them pass the responsibility to the authority, who offers the option, and therefore take it without much hesitation (Campbell-Arvai *et al.* 2014). In line with the in-group conformity, another popular way of nudge theory is the social-proof heuristic method: an option is presented with the information that it is the preference of the majority, which would lead customers to follow the trend and make the desired choice (Cheung *et al.* 2017). Moreover, nudge theory also includes recommender systems and more intelligent algorithms to personalize an environment for users to nudge them to choose the desired option gently (Mareike 2021). In a word, when we act within groups, we tend to take the environment and others' behavior as the references. Thus, the in-group influence, derived from the shaped information and responsibility, could be used as a tool to alter the environment in a gentle way and nudge people to lead a more sustainable life.

3.1.4 *The personal touch: An individual perspective on ecological transition*

Given this general overview on how cultural and social dynamics shape people's minds and beliefs, we can now narrow down our perspective on how individuals think and feel about climate change and the issues it generates. In order to delve into the problem, we can orchestrate our work around two simple and almost trivial questions. Given the fact that almost everybody acknowledges the urgency of taking a stand regarding the climate crisis, why is it that we do not act green, even if we knew how it would benefit both the planet and us? Secondly, once we gain consciousness on the matter, what can we do to try and solve the problem?

3.1.4.1   Avoiding the anxiety

One of the main reasons, why we avoid dealing with climate change can be found in the anxiety the situation triggers (Clayton *et al.* 2017). We reached a point so serious that the anxiety coming from the climate crisis has become a public issue, coming to the attention of doctors and psychologists around the world, as mentioned above, leading to the definition of *eco-anxiety*, given by Albrecht (2011) and others. Nevertheless, it seems like we are not making an effort to lessen its effects, living our lives without worrying about the issue: this is due to what psychologists usually refer to as *psychological distance* (Wang *et al.* 2019), meaning that in order to avoid bad emotions and excessive worries, people tend to think of it as something far away in space and in time or, at least, from themselves. In other words: "Although most people are generally aware that climate change is occurring, it continues to seem distant: something that will happen to others, in another place, at some unspecified future date (McDonald *et al.* 2015). Terms such as climate change and global warming draw attention to the global scale rather than the personal impacts" (Rudiak-Gould 2013). It seems also possible to blame, in part, even the media: they "have been criticized for promoting an inaccurate perception of climate change (Antilla 2005): for example, that there is more scientific controversy about climate change than actually exists" (Clayton *et al.* 2017).

However, closing our eyes and looking on the other side trying to avoid anxiety does not solve any problem: if anything, it risks making things worse. Therefore, it seems like a change in perspective is needed, a change involving also the semantic we use to address this crisis: we are not trying to save the planet, we are trying to save ourselves and our way of living. To face this crisis and win this battle, people have to develop the psychological tools and resources needed to cope with our own emotions. To use the words of Antonino Ferro (2007a), people ought to "*Live* the emotions" instead of "*avoiding* the emotions": sometimes it can be scary and overwhelming, but when it comes down to climate change denying its effect on the people can lead nowhere.

3.1.4.2 Effort and sacrifices: Evaluating the benefits of acting greener

Another big issue of living a greener lifestyle is how much effort it requires: going by bicycle instead of using the car, buying more expensive items at a grocery store, renouncing tastier food to become vegan, recycling, avoiding wasting, avoiding buying stuff that we do not need or taking the plane, it always seems like living green requires a lot of sacrifice and to renounce a lot of comfort, resulting in feeling like there are very little benefits coming from it. But is it like that?

Unfortunately, it is true that living green requires lessening our comforts, many tasks are without any doubt, more difficult than living without caring for our carbon footprint. Nevertheless, if we truly look at all the benefits they give back to us, the perspective radically changes, resulting in the benefits, largely paying off the costs we have to undertake (Leaman 2015). Using wind or solar energy, apart from being extremely ecological (Kalogirou 2004), in the long run leads to energy autonomy for each household that uses them, and much smaller expenses for, for instance, heating or electricity. Most likely, using a bike instead of a car, in the long run, leads to a healthier lifestyle, a better shape; it reduces the risk of cardiovascular accidents and diseases and, last but not least, costs far less than using a car.

Therefore, what can be done is focusing a little bit more on the long term instead of the short term, trying to highlight all the benefits coming from living in a green way and trying to give some little benefits even on a short run (such as benefits for using a bike, tax reductions if reusable energy supplies are used, ensure lower prices also for ecological products at grocery stores and supermarkets...), like it is done in many countries (Leaman 2015). By doing so, it can be hoped that the decision-making process of individuals will lean towards better choices and a greener lifestyle, more accessible to anyone.

Moreover, by encouraging people with some benefits, and also by avoiding that they choose the easiest way, another occurrence can be avoided: the so-called *Cognitive Dissonance*. Cognitive Dissonance is a psychological phenomenon (Festinger 1957) by which people try to obtain more coherence between what they think and what they do by reshaping their own ideas *after* having done something that is against their own beliefs, since inconsistency among beliefs or behaviors causes an uncomfortable psychological tension. To better explain its functioning, in an experiment proposed by Festinger (1955) participants were asked to undergo a terribly boring experiment and then lie to the next participant (actually an experimental accomplice) about the task, telling them it was more enjoyable than it really was. Some people were paid $1 to do so, others $20. Participants were then asked to rate the experimental tasks, describing it as enjoyable or not. Festinger correctly predicted that people who were paid $1 evaluated the task as more enjoyable than people getting paid $20, since they had a little compensation for their time and their lies, resulting in a cognitive dissonance. The same cognitive process could be thought to happen every time we act without caring about the environment or our carbon footprint: the result would be to lessen the expected effects of our action on the environment or even denying the urge of the climate crisis we are living in, whereas to be pushed towards a greener lifestyle could use cognitive dissonance as a tool to our cause (e.g. eating less tasty foods but evaluating them as tastier than they are).

Lastly, we have to acknowledge that the common thought that everything will be sorted out and we will all be fine, it is largely grounded in a sense of omnipotence that humans tend to have and in which they seek refuge when anxiety or fear overpower them: that does not mean that it is true, even if they are tools humankind developed throughout the years thanks to its evolution (Del Giudice 2018). We have to remember that even if we have a huge power over nature and we are indeed able to adapt part of the environment in order for it to fit and meet our needs, we are the ones who mostly have to adapt to the outside world. We are not on the top of the ecosystem, we are part of it, and to preserve one means to preserve the other.

3.1.4.3   What to do: Empowerment and implementing effective policies

So, what can we do to make people act greener and in a more responsible way? First of all, the climate crisis we are facing has to be described as the urgent issue it is, without minimizing it and without avoiding addressing the problem. In order to facilitate this process, the acceptance process ought to be enhanced, giving the people the right instruments and resources to face and elaborate the anxiety that comes from this crisis. This can also mean to strengthen the psychological support the general population can receive, but also make people feel empowered and with an active role in solving the problem.

Thus, firstly, enhancing agency could be a good strategy to make people feel responsible for their own actions, giving them the possibility to make better choices: especially taking into account the fact that many who are trying to reach a greener lifestyle are not able to see the results coming from their behavior straight away, reducing their perception of making a proper difference considerably: over 80% of young people feel like we reached the point of no return and no good future can be. However, it is important to stress that individuals are not to be blamed too much: apart from not being right, it could be counterproductive. Many people are not in the proper conditions to enact such a change in their behaviors, and there would be no point in blaming them: it could also lead to them becoming a scapegoat. It is a precarious balance: people should feel active, responsible, empowered and conscious of the reach of their behaviors, believing in their power to change the world and the harm coming from acting otherwise; nevertheless, an excessive blame ought to be avoided, as well as sit back and, thanks to the responsibility diffusion, not even feeling responsible for the damage dealt to the world.

Secondly, it would be helpful to implement some proper and effective policies to grant that the efforts of individuals would not be in vain and to facilitate a greener lifestyle. In this sense, large corporations, especially the ones more responsible for pollution, should be made responsible for the situation they largely contribute to worsen (The Guardian). One could even advocate that by making large corporations greener, a great deal of the problem would be solved, and it would also positively impact individuals, making them feel like their sacrifices are worthy of the cause and will lead to some results.

3.1.5 *The need of empathy*

To make a choice in favor of the greener lifestyle always means to think about the advantages of society, not one's own benefits. What drives people to care for the needs of others and cooperate? Humans, like other primates, are highly social creatures, and cooperation with congeners should be a particularly desirable form of social behavior. So, we must have some "built-in" mechanisms to provide this kind of behavior. Actually, not only primates have inborn or genetic mechanisms providing social behaviour; for example, kin selection is thought to be an explanation for altruism in a wide range of animals, including insects (Foster *et al.* 2006). Another kind of social behaviour observed in some mammals is reciprocal altruism. However, animals with more complex social interactions like primates need a mechanism that would enable them to perceive and be sensitive to the emotional states of others, coupled with a motivation to care for their well-being, and empathy is thought to be such mechanism (Decety *et al.* 2016). Both in animals and humans, the ability to empathize mediates prosocial behaviour. Biological basis of empathy can be the oxytocin- or arginine vasopressin-modulated neural circuits related to empathic behavior (Decety *et al.* 2016) and mirror neurons, that can be found in primates and thought to be the basis for human empathy (Haeusser 2012).

The construct of empathy can be divided into cognitive and affective dimensions (Walter 2012, Gonzalez-Liencres *et al.* 2013). They are thought to have different neuronal basics. Thus, fMRI studies have reported limbic structures such as the amygdala, the anterior insula, and the anterior cingulate cortex to be part of the neural bases of affective empathy (Lamm *et al.* 2011, Bernhardt and Singer 2012, Gonzalez-Liencres *et al.* 2013). The prefrontal cortex, including dorsolateral, ventromedial, and orbitofrontal regions, is supposed to be related to cognitive empathy (Van Overwalle and Baetens 2009, Bernhardt and Singer 2012).

The model of the empathic act is proposed by Stein (1970). The author describes three stages of empathy: on the first stage, you listen carefully to the other person trying to get yourself in their place. At the second stage, you feel identification and cross over of self, which just happens to you, it is not what you do deliberately. And the third stage brings your self back to you and it is where you feel sympathy for the other. So only the first and the third stages of empathy are under voluntary cognitive control. Does that mean that these steps can be taught?

Lam and colleagues (2011) describe a number of methods to teach empathy, like for example, experiential, when the instructors design experiences for trainees, skill training, like a skill of active listening, video stimulus, when you watch someone's empathic behaviour, writing, when you actively put yourself in the place of someone who needs being treated with empathy, and mindfulness training. The last technique enables to facilitate the emotional component of empathy — the second stage, the feeling of dissolving the barriers between yourself and others.

Another essential issue concerning empathy is how to develop this ability in children. McDonald and Messinger (2011) discuss factors that facilitate the development of the ability to empathize in children. They are within-child contributions such as genetics, neural development, and temperament, and socialization factors including facial mimicry and imitation, parenting, and parent-child relationships (or adult-child relationship, in general). There are a number of techniques, parents and educators may use to facilitate empathy in children, such as, for example, training in interpersonal perception and empathetic responding, focus on similarities between oneself and others, exposure to emotionally arousing stimuli, etc. (Cotton 1992).

Empathy underlies not only pro-social, but also pro-ecological behaviour, so it is important to develop empathy in children in the human/nature relationship. In the book "Environmental communication for children", Hawley (2022) raises the question, whether mediated experiences with nature can contribute to the development of environmental sensitivity and nature-connectedness. What he proposes is that the animated, digital, and virtual experiences such as for examples movies, cartoons like Wall-e or videogames like Minecraft open a space for reconfiguration of the human/nature relationship. Thus, empathy can be one of the mechanisms of pro-ecological behaviour, and we can develop the ability to empathize both in children and adults.

3.1.6 *Conclusions*

Trying to find a solution to a problem like climate change requires world-wide action and effort. For it to succeed, strong and structural changes are needed: individuals' behaviors and beliefs, social norms and habits and cultural aspects ought to steer towards greener and more sustainable alternatives, leading to more sustainable lifestyles. To do so, a strong will is required, since there are no immediate benefits coming from these changes. Nevertheless, no result can be achieved without effort and a little bit of sacrifice, and the benefits coming from them are unfathomable. Without blaming individuals too much, who have no resources to make good choices, we have to understand that the time has come when we have little or no choice: trying to choose a greener lifestyle is now necessary for our survival and for keeping our world as beautiful and safe as it is. Achieving a greener lifestyle is in reach, and it requires much less effort than we imagine: it is just a matter of changing our perspective on the issue and choosing sustainable options.

References

Albrecht, G. Chronic environmental change: Emerging "psychoterratic" syndromes. In I. Weissbecker (Ed.), Climate change and human well-being: Global challenges and opportunities, pp. 43–56. New York, NY: Springer (2011).

Antilla, L. Climate of skepticism: US newspaper coverage of the science of climate change. *Global Environmental Change* 15, 338–352 (2005).

Armin, F. and Szech, N. Diffusion of being pivotal and immoral outcomes. Working Paper Series in Economics, *Karlsruher Institut für Technologie* (KIT) (2017).

Bernhardt, B. C., and Singer, T. The neural basis of empathy. *Annu. Rev. Neurosci.* 35: 1–23 (2012).

Bischoff, F. and Deij, V. *Our eco-anxious Future*, University of the Arts, London (2022).

Boorman, E.D., O'Doherty, J.P., Adolphs, R. and Rangel, A. The behavioral and neural mechanisms underlying the tracking of expertise. *Neuron* 80, no. 6: 1558–1571 (2013).

Campbell-Arvai, V., Arvai, J. and Kalof, L. Motivating sustainable food choices: the role of nudges, value orientation, and information provision. Environment and Behavior 46, no. 4: 453–475 (2014).

Cheung, T., Kroese, F., Fennis, B. and de Ridder, D. The hunger games: using hunger to promote healthy choices in self-control conflicts. *Appetite.* 116: 401–409 (2017).

Clayton, S., Manning, C. M., Krygsman, K. and Speiser, M. *Mental Health and Our Changing Climate: Impacts, Implications, and Guidance.* Washington, D.C.: American Psychological Association, and ecoAmerica (2017).

Cotton, K. *Developing empathy in children and youth.* Northwest Regional Educational Laboratory (1992).

Decety, J., Bartal, IB-A., Uzefovsky, F. and Knafo-Noam, A. Empathy as a driver of prosocial behaviour: highly conserved neurobehavioural mechanisms across species. *Phil. Trans. R. Soc. B* 371: 20150077 (2016).

Del Giudice, M. *Evolutionary Psychopathology: a Unified Approach.* New York: Oxford University Press (2018).

Donald, J. N., Sahdra, B. K. *et al.* "Does your mindfulness benefit others? A systematic review and meta-analysis of the link between mindfulness and prosocial behavior". *British Journal of Psychology* 110, no. 1, 101–125 (2019).

El Zein, M., Bahrami, B. and Hertwig, R. Shared responsibility in collective decisions. *Nat. Hum. Behav.* 3: 554–559 (2019).

Ferro, A. Evitare le emozioni, vivere le emozioni. Milano: Raffaello Cortina (2007a).

Freeman, J. B., Schiller, D., Rule, N. O. and Ambady, N. The neural origins of superficial and individuated judgments about in-group and out-group members. *Hum. Brain Mapp.* 31: 150–159 (2010).

Foster, K. R., Wenseleers, T. and Ratnieks, F. L. Kin selection is the key to altruism. *Trends in ecology & evolution.* 21, no. 2, 57–60 (2006).

Gregory, A. *Eco-anxiety: fear of environmental doom weighs on young people.* [online] Available at: https://www.theguardian.com/society/2021/oct/06/eco-anxiety-fear-of-environmental-doom-weighs-on-young-people (Last access: 21.06.2022).

Gomez-Uribe, C.A. and Hunt, N. The Netflix Recommender System. *ACM Transactions on Management Information Systems* **6**, no. 4, 1–19 (2015).

Gonzalez-Liencres, C., Shamay-Tsoory, S. G. and Brüne, M. Towards a neuroscience of empathy: ontogeny, phylogeny, brain mechanisms, context and psychopathology. *Neurosci. Biobehav. Rev.* 37: 1537–1548 (2013).

Haeusser, L. F. Empathy and mirror neurons. A view on contemporary neuropsychological empathy research. *Praxis der Kinderpsychologie und Kinderpsychiatrie* 61.5: 322–335 (2012).

Hawley, E. Environmental Communication for Children: Media, Young Audiences, and the More-Than-Human World. Springer Nature (2022).

Hayashida, K., Yu, M., Nishi, Y. and Shu, M. Changes of Causal Attribution by a Co-actor in Situations of Obvious Causality. *Frontiers in Psychology* 11 (2021).

Heatherton, T. F. Neuroscience of self and self-regulation. *Annu. Rev. Psychol.* 62: 363–390 (2011).

Hoekstra, R. *The Knowledge Reengineering Bottleneck, Semantic Web Interoperability, Usability, Applicability.* IOS Press (2010).

Iberdrola. *Eco-anxiety: the psychological aftermath of the climate crisis.* [online] Available at: https://www.iberdrola.com/social-commitment/what-is-ecoanxiety (Last access: 19.06.2022).

Irlenbusch, B. and Saxler, D.J. The role of social information, market framing, and diffusion of responsibility as determinants of socially responsible behavior. *Journal of Behavioral and Experimental Economics* 80: 141–169 (2019).

Kalogirou, S. Environmental benefits of domestic solar energy systems, *Energy Conversion and Management,* Volume 45, Nos. 18–19, pp. 3075–3092, https://doi.org/10.1016/j.enconman.2003.12.019.

Lamm, C., Decety, J. and Singer, T. Meta-analytic evidence for common and distinct neural networks associated with directly experienced pain and empathy for pain. *Neuroimage* 54: 2492–2502 (2011).

Leaman, C. The benefits of solar energy, *Renewable Energy Focus,* Volume 16, Nos. 5–6, pp. 113–115, https://doi.org/10.1016/j.ref.2015.10.002.

Li, P., Jia, S. T., Liu, Q., Suo, T. and Li, H. The influence of the diffusion of responsibility effect on outcome evaluations: electrophysiological evidence from an ERP study. *Neuroimage* 52: 1727–1733 (2010).

Lin, C., Liu, D., Tong, H. and Xiao, Y. Spiral of Silence and Its Application in Recommender Systems. *IEEE Transactions on Knowledge and Data Engineering* 34, no. 6: 2934–2947 (2022).

Marks, J., Copland, E., Loh, E., Sunstein, C.R. and Sharot, T. Epistemic spillovers: Learning others' political views reduces the ability to assess and use their expertise in nonpolitical domains. *Cognition* 188: 74–84 (2019).

Masterson, V. A. *et al.* The contribution of sense of place to social-ecological systems research: a review and research agenda. *Ecology and Society* 22.1 (2017).

Matthes, J. Observing the "spiral" in the spiral of silence. Int. J. *Public Opinion Res* 27, no. 2: 155–176 (2015).

McDonald, N. M. and Messinger, D. S. The development of empathy: How, when, and why. *Moral Behavior and Free Will: A Neurobiological and Philosophical Approach* 333–359 (2011).

McDonald, R., Chai, H. and Newell, B. Personal experience and the 'psychological distance' of climate change: An integrative review. *Journal of Environmental Psychology* 44, 109–118 (2015).

Mezulis, A. H., Abramson, L. Y., Hyde, J. S. and Hankin, B. L. Is there a universal positivity bias in attributions? A meta-analytic review of individual, developmental, and cultural differences in the self-serving attributional bias. *Psychol. Bull.* 130: 711–747 (2004).

Milgram, S. Behavioral study of obedience. *Journal of Social Psychology*, 72:207–217 (1963).

Möhlmann, M. Algorithmic Nudges Don't Have to Be Unethical. *Harvard Business Review* (2021).

Pelzer, P. *Bicycling as a Way of Life: A Comparatie Case Study of Bicycle Culture in Portland, OR and Amsterdam.* Oxford (2010).

Rudiak-Gould, P. *'We Have Seen It with Our Own Eyes': Why We Disagree about Climate Change Visibility.* Wollmence, Academia (2013).

Schensul, J.J. Community, Culture and Sustainability in Multilevel Dynamic Systems Intervention Science. Am J Community Psychol 43, 241–245, DOI 10.1007/s10464-009-9228-x (2009).

Stein E. On the Problem of Empathy. 2nd ed. The Hague, the Netherlands: Martinus NijhoffDr W Junk Publishers (1970).

Sunstein, C.R. *Infotopia: How many minds produce knowledge.* Oxford University Press (2006).

Stallen, M., Smidts, A. and Sanfey, A. Peer influence: neural mechanisms underlying in-group conformity. *Frontiers in Human Neuroscience* 7: 1662–5161 (2013).

Tagliabue, M. and Carsta, S. Feeding the behavioral revolution: Contributions of behavior analysis to nudging and vice versa. *Journal of Behavioral Economics for Policy* 2, no. 1, 91–97 (2018).

Tajfel, H., Billig, M. G., Bundy, R. P. and Flament, C. Social categorization and intergroup behaviour. *Eur. J. Soc. Psychol.* 1:149–178 (1971).

The Guardian: https://www.theguardian.com/sustainable-business/2017/jul/10/100-fossil-fuel-companies-investors-responsible-71-global-emissions-cdp-study-climate-change.

Van Overwalle, F. and Baetens, K. Understanding others' actions and goals by mirror and mentalizing systems: a meta-analysis. *Neuroimage* 48: 564–584 (2009).

Walter, H. Social cognitive neuroscience of empathy: concepts, circuits, and genes. *Emot. Rev.* 4: 9–17 (2012).

Wang, S., Hurlstone, M.J., Leviston, Z., Walker, I. and Lawrence C. Climate Change From a Distance: An Analysis of Construal Level and Psychological Distance From Climate Change. *Front Psychol.* (Feb 22, 2019); 10:230. doi: 10.3389/fpsyg.2019.00230. PMID: 30853924; PMCID: PMC6395381.

Yates, J.F. and de Oliveira, Stephanie. Culture and Decision Making. *Organizational Behaviour and Human Decision Process* 136, 106–118 (2016).

Zaki, J., Schirmer, J. and Mitchell, J. P. Social influence modulates the neural computation of value. *Psychol. Sci.* 22: 894–900 (2011).

4 From theory to practice: Bicycle as a main transportation mode

Thomas Carey

University of Leeds, University of Cambridge, UK

Yibo Cao

School of Psychological and Cognitive Sciences, Peking University, China

Elizaveta Baranova-Parfenova

Graduate Institute of Mind, Brain and Consciousness, Taipei Medical University

4.1 Abstract

The modern world has many problems never before faced in the history of humanity, including climate change. One of the most crucial contributors to this catastrophe is the carbon dioxide (CO_2) greenhouse gas. Breaking down the total emissions into sectors shows that the biggest amount of CO_2 emissions is released by several areas, such as heating, energy and transportation. A variety of actions toward a zero-emissions future are meant to be implemented on individual, industrial and governmental levels. Thus, in 2015 the Paris Agreement was signed, with reducing greenhouse gasses as one of the goals set out therein. It is important to note that there are no universal solutions for all societal structures and places, and that is why one needs to create and execute locally. Nonetheless, there are particular cells of the vast majority of societies that might accept general ideas about sustainable transportation. In our project, we want to check the possibilities to improve the transportation situation of university students. Some people do care about the environment and follow their beliefs, however, others have no intention or resources to do so. We argue that making a default mode as the most environmentally-friendly option might be a good way to promote sustainable behavior. It is a common observation that riding a bicycle becomes less convenient during precipitation and poor weather conditions such as rain, snow or hail. Our suggestion is to implement roofs on bicycle routes, especially on the ones leading from the dormitories to the university buildings. There are some existing suggestions about sheltering systems for such types of traveling, nevertheless, they seem not to be implemented in real-world situations. Via modeling and open data from bicycle rent systems, we want to show that particular roads are less busy during rainy weather. Thus, roofs above these routes might be a solution to promote a bicycle as the main mode of transportation despite the time of the year. Finally, we want to introduce additional details of such a sheltering system, which might make a real contribution to the sustainability of students' transportation.

4.1.1 *Introduction*

Climate change is a huge threat to the Earth's ecosystem and humanity. It is admitted that the main cause of current rapid climate change and other environmental issues happens because of human activity (Stern 1992), such as emissions of carbon dioxide (CO_2). Two of the sectors emitting such greenhouse gasses are the energy and transport sectors. Transportation is a huge contributor to CO_2 levels in the atmosphere. Two types of solutions are suggested to solve this issue: top-down (governmental and corporate levels) and top-up (individual level). On the level of the governments, the Paris Agreement was signed with a goal to reduce the amount of greenhouse gases emissions. Since the top-down approach is not directly accessible to many researchers, they focus their studies on the individual level.

4.1.2 *Pro-environmental behavior: Phenomena and obstacles*

An individual level of behavior with an intention toward lowering an individual ecological impact or reducing the amount of using resources can be called pro-environmental behavior (PEB) (Lange and Dewitte

2019). It includes a spectrum of behaviors that lead to decreased harm to the environment. Some examples of such behaviors are about "making something", while others are in the category of "avoiding doing something" (Mackay and Schmitt 2019). There are a variety of approaches to the promotion of PEB. Some people constantly think about the impact of their actions on the environment and follow their beliefs, while others do not see reasons to behave more sustainably or do not have the mental or physical resources to do so. Additionally, not all people match their behavior to their beliefs.

There are many obstacles that do not ease the PEB execution, such as the rebound effect (Kratschmann and Dütschke 2021). Here, a person implements a new habit or technology which leads to decreased environmental impact. Because of such a reduction in effect, one can think that their environmental action is fulfilled, leading to additional activity to spend the leftover resources. Thus, the rebound effect shows that a conscious reduction of harm may lead to additional behavior which not only damages the overall impact but increases it.

Another great example of psychological obstacles is the negative footprint effect (Gorissen and Weijters 2016). It was shown that people do not sum up the effect of a bundle of items, but average it. So, adding an "environmentally friendly" item to the other item makes one think that the bundle's impact is less compared to the initial "non-green" item alone.

Thus, it is important to help people make an environmental choice with the help of the architecture of their environment. In an ideal situation, the most environmentally desirable option should be the most convenient one. The prominent examples of behavior architecture are nudging and default mode.

Nudging is a way to guide a person towards a desired behavior (Schubert 2017). In the context of PEB the goal of nudging is to give an idea what behavior is the most desirable. For example, "The Little Book of Green Nudges" (De Luigi *et al.* 2020) gives lots of suggestions on how to change the environment in a way to provoke desired behavior without changing the amount of options. For example, in the book the authors suggest installing free self-repair bicycle stations in a territory of university campuses to provoke more exploitation of bicycles.

Another way to construct a desired behavior is a default mode condition (Pichert and Katsikopoulos 2008). In this case, in situations of choice, the default mode should be the most environmental one. Sometimes the default option would mean that the individual not only should take the sustainable alternative, but avoid a purchase or service at all (such as avoiding all the packaging instead of choosing a recyclable one). The default options help to reach desired behavioral goals without changing the value systems of people who do not want or do not have an opportunity to implement PEB. Also, default mode options require the companies and administration to do more research and take more responsibility to change some environment (for example, infrustructure) or arrange a new one.

4.1.3 *Environmental and economic benefits of cycling*

It is clear that high cycling culture brings significant environmental benefits. As an example, the project Drawdown (Houthuijs *et al.* 2014) estimated that with increased amount of cyclists, the CO_2 emissions can be reduced up to 4.63 gigatons by the year 2050. The economical benefits of cycling are also estimated and seem to be very prominent. For example, it was estimated that an increased number of bicycle users will lead to more than 150 billion euros of total annual benefit (Kratschmann and Dütschke 2021). Also, only for Europe, 400,000 jobs related to cycling can be created (Blondiau *et al.* 2016).

It is important to note that there are hidden environmental costs of road exploitation (Kratschmann and Dütschke 2021), highlighting the pros of light weight of a bicycle. It was shown that not only the amount of resources should be taken into consideration when we compare a bicycle and a car, but also their weight since it affects the durability of road usage before repair.

Talking about the other benefits, it was shown that increased amount of cyclists increases the safety of the road to all the road users (Marshall and Ferenchak 2019). Additional non-obvious benefits of cycling are reduction of premature death and decreased noise pollution (Houthuijs *et al.* 2014, Hamilton and Wichman 2018).

Of course, with all the benefits there are obvious obstacles, such as time consumption (Wang *et al.* 2020). It was estimated that 1 kg of CO_2 saved by using a bicycle leads to 0.14 lost hours (8 lost minutes) of time. On the one hand, it seems that losing time is worth the benefits. On the other hand, it would be better if people from all economic classes could choose bicycles as a main transportation mode when they have an opportunity to make such a choice.

4.1.4 *Hidden costs of using cars*

The most abundantly used form of private transportation, the automobile, has many negative impacts to the environment, but also to the people who drive them and wider society. For the environment, the exhaust fumes pollute the atmosphere. Electric cars, which are becoming more common in some countries, are better in this regard, but still have many of the other issues and are not fully sustainable yet. Building roads often damages wild habitats, and the vehicles themselves are incredibly dangerous to animal wildlife which is not prepared evolutionarily for such a predator. For example, the hedgehog's natural defense mechanism is to roll itself into a ball, which is effective against predatory animals but not against cars. Roads are also often built through lower-income communities, displacing these people who already struggle to find stability in their lives. As well as animals, humans are also at great risk from motor vehicle deaths, with one person losing their life to a traffic collision roughly every 10 minutes (Marshall and Ferenchak 2019). This makes traveling by car significantly more dangerous than trains, airplanes or any other common form of transportation other than the motorcycle, but often it is pedestrians and cyclists who are collateral damage in these incidents. In the US, the leading cause of death for children and young adults is car accidents (Marshall and Ferenchak 2019).

Cars also carry a significant financial burden on both their owners and the people who live in car-centric societies. Cars depreciate in value quickly, with a car being worth around 20% less every year (Pichert and Katsikopoulos 2008), so selling a car generally returns people a small fraction of what they paid for it. There are frequent costs for insurance and maintenance, and as incidents that require paying for this can occur without warning, this can be very dangerous to people from households with low income. Also, road maintenance is in most societies paid for by the taxpayer, including those who do not drive — the damage done to a road by a vehicle is proportional to the vehicle's speed and to the fourth power of its weight, meaning that one journey in an average car does as much damage as over 17 thousand bicycle journeys, and one journey in a 9 ton (9000 kg) truck does as much damage as 7 million bicycle journeys (Schubert 2017). The manufacturing of private cars also produces a large amount of CO_2, and often uses certain elements like palladium and platinum that are more harmful to the environment. At the same time, production of bicycles does not demand these (Stern 1992).

Car centric society also isolates people. When people live in suburban areas with no amenities or communal areas within walking distance, and no suitable public transport system exists, their only option is to drive an individual automobile. This means that they rarely have any social interaction during their commute, as instead of bumping into someone on the walk to the shops their only interconnection with other cars tends to be when annoyed at their driving or traffic in general. This is a problem, especially for children, because they cannot go and see their friends outside of school or other activities, and lack of ability to go places impairs them physically and mentally (Wang *et al.* 2020). While children often walk or take public transport in societies where this is possible, it has become such an alien concept in suburbia (especially in the US) that parents have been reported to the police for letting their children walk distances of under 500 meters. Because of this, they do not learn independence as well. As previously mentioned, noise pollution is another issue — it is unpleasant for those living around it, and can lead to issues like hearing loss — as well as potentially harm unborn babies and lead to premature deaths.

4.1.5 *Cycling culture*

Since bicycles are so important to the sustainable development of the world, and people can save energy, reduce emissions and stay healthy at the same time, it seems like a perfect choice for short-distance travel.

What are the obstacles of choosing this green and healthy way to travel? Is it because the number of cars far exceeds the number of bicycles? Chen with colleguages suggests an answer (Chen *et al.* 2022). In this study they show that global production of bicycles increased from 20.7 million units in 1962 to 123.3 million units in 2015 with an annual growth rate of 3.4% which is higher than global car production in the same period. Bicycles seem to be produced at least no less than cars. The authors point out that ownership of bicycles plays a less important role in their usage, which means higher bicycle ownership does not necessarily guarantee increased bicycle use.

As a country of cycling, Denmark's cycling model is worth learning. In Denmark, nine out of ten people own a bike, and they cycle 1.6 km a day on average. Cycling accounts for a quarter of all personal transport in Denmark for distances of less than five kilometers (Lange and Dewitte 2019).

Additionally, bike lanes are painted with bright colors as a reminder to reduce the interference of motor vehicles, and some lanes are raised to be higher than the roads to ensure the safety of cyclists. Furthermore, there are 45 degree trash cans made specifically for cyclists.

According to Chen's study, if people follow the Danish pattern of cycling, which is on average 1.6 km a day, they could save 414 million tons of carbon emissions a year, equivalent to the UK's total emissions in 2015. If everyone were to follow the Dutch cycling model, it would save 686 million tons of carbon a year.

In Denmark, bicycles are widely used for exercise, leisure and entertainment. Denmark has also made life easier for cyclists in many ways. The landscape of this country is flat and the government encourages cycling. In another place with developed cycling culture, the Netherlands, there is a national network of bicycle routes, each of them having a color-coded cycle path. Certain crossroads feature dedicated traffic signals for bicycles, and some nature reserves are inaccessible to motor vehicles and have special bike lanes.

In this fast-paced era of efficiency and speed, it is hard to slow down and enjoy cycling, and the number of people riding has plummeted. But the Danish capital still has a strong cycling culture. According to Copenhagen's official statistics, the number of cyclists in Copenhagen reached 50% of the city's population in 2015. How did Copenhagen manage to promote cycling as a green, environmentally friendly and sustainable way to travel?

The first thing is the urban planning of Copenhagen. In addition to the special traffic lights for bicycles at some intersections, Copenhagen has bike stops 5 meters ahead of cars at intersections, allowing cyclists to pass the intersection first. In some high-traffic streets, bike lanes will be as wide as motor vehicles to improve the capacity and speed of bicycles (张长征 2015).

Second, Copenhagen designed the "green wave", which guides cyclists along the road at a speed of 20 km/hour by flashing green lights on short posts. If cyclists keep their speed in line with the flashing lights, they can go all the way because there will be no red light (Mackay and Schmitt 2019).

4.1.6 *Case study*

Based on the reviewed theoretical implications, we decided to look into the real-life behavior of cyclists during unpleasant weather conditions. In this study we used a case from Cambridge, UK, which is by a large margin the most popular city for cycling in the UK, with a third of residents cycling at least three times a week. It is very flat and relatively small, having been built around the university, and has over 80 miles (128 km) of dedicated cycle routes (Blondiau *et al.* 2016). Also, there are various schemes to restrict driving in the historic center, with its narrow and busy roads, such as parking restrictions and cyclist priority roads, which encourage cycling. Because of this, it has a strong cycling culture. There are numerous "cycle counters" around the city and surrounding towns, which have data that update in real time when a bicycle goes past and is displayed publicly on these devices. The data is publicly available online (Chen *et al.* 2022). Our goal was to see just how much change in weather would affect how many people cycle past a certain point. It was expected that people would be less likely to cycle if the precipitation was stronger, but we wanted to obtain some quantitative data.

Because Cambridge is such a student-oriented city, there are huge disparities between the number of cyclists in the city center during term time and student holidays. Additionally, with examinations and other events in the student calendar affecting the amount of people cycling far more than any environment-based factors, picking a location in the center of the city would not have yielded useful data. Because of this, we chose a location far enough outside of the city center to be somewhere that is rarely visited by students, but still close enough that it has a large number of cyclists, many of which are traveling to and from the city center of Cambridge. The suburban area of Newtown, to the south-west of the city center, was an ideal location for this.

There are many online sources for historical weather data, but we quickly found that a large number of them were not usable for our purposes. Some of the sources only showed one summary for each day, which does not give a particularly precise view of how much it rained on any given particular day, while other sources required expensive subscriptions to download data. The data at (De Luigi *et al.* 2020) proved very useful, with a detailed breakdown of the weather in every half-hour segment of every day. For each of the 365 days in 2018, we counted how many of the half-hour intervals between 6am and 9pm (as not many people are likely to cycle during the night) had: no precipitation, showers, light precipitation, moderate precipitation or heavy precipitation. While we only looked at precipitation, it is worth noting that the more precipitation there is, the more likely it

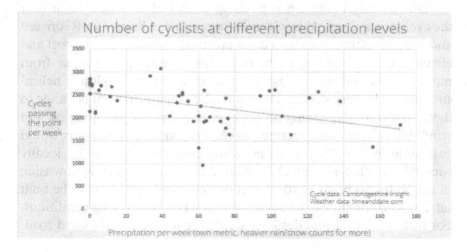

is that other weather factors would be discouraging to cyclists — for example, low temperature, low light level and slippery ground, all of which are heavily correlated with levels of rain and snow.

We assigned a value to each of these five levels — there were a few sensible ways of doing this, but the one we ended up choosing was 10 for heavy precipitation, 6 for moderate, 3 for light, 1 for showers and 0 for no precipitation. It should be noted that other assignments, such as 4, 3, 2, 1, 0, gave very similar results. From this, we plotted the number of cycles registered on the cycle counter each week with the sum of the precipitation values across the week and obtained the following graph.

There is a clearly visible negative correlation, with a Pearson correlation coefficient of −0.4684 — a moderate negative correlation. The y-intercept of this graph (the expected number of bicycles passing the point during weeks with no precipitation at all, which happened several times during the summer heatwave that year) is 2551. If this many cyclists were to pass the cycle counter every week of 2018, it would mean an increase of 10.71%. Solutions that aim to lessen the effects of precipitation will not be able to achieve the full amount of this increase, as people are much more likely to go for a leisurely bike ride when the weather is pleasant, which is difficult to simulate in a cost-effective way. But by reducing the annoyance of cycling during precipitation and associated bad weather, a portion of this increase could be realized.

One potential solution is building a kind of roofing system over the cycle lanes. Inspired by German organization Bambutec (Gorissen and Weijters 2016), who have done practical research into the cost and effectiveness of building such roofs. They are simple to make from mostly wood and do not use a lot of material but exploit a helical structure to be very strong. There are adjustments that can be made to these structures depending on the expected level of precipitation and wind, but the general structure attached to existing cycle paths would cost as little as €300 per meter, while creating them out of locally sourced wood helps save CO_2 compared to similar designs. A downside of this is that it requires some sort of dedicated cycle lane to be built on, but adjustments can be made depending on the existing conditions. For instance, one could design a similar system over a standard road, by attaching the roofs to existing lamp posts for support, with gutters to funnel water into drains, and adding LED light strips. These would prevent precipitation from hitting the cyclists and the road surface on which they cycle, as well as improving visibility. Any solution will be more effective if it takes into account the existing infrastructure and uses that to its advantage as best as possible.

Another example of adapting existing, possibly disused, infrastructure into something that promotes environmentally friendly transport is that of greenways. Greenways are paths that once served an industrial purpose (such as railway lines and canal towpaths) that have been adapted for walking, cycling and similar non-motorised means of transport. Due to their initial purpose, these are often very flat which makes for a more pleasant and accessible journey, and the sides have been reclaimed by nature which provides a habitat for wildlife as well as a nicer view for the users of the greenway. Often, they are maintained in a similar way to parks. There are a variety of these across the world, which often make use of natural features such as rivers. One successful example is the Waterford Greenway in south-east Ireland, which opened in March 2017 and was visited by a quarter of a million people by the end of that year — more than double the population of County Waterford in which it is located (Hamilton and Wichman 2018). This particular greenway is 46 km long and has sections on the coast, sections along the river Suir and a preserved railway along a small section of the greenway, as well as numerous cafes and similar businesses for the people who travel along it.

4.1.7 *Discussion*

(Elizaveta Baranova-Parfenova) In the current text we investigated the conditions and possible solutions of transportation mode of the individuals. We show that humans have a variety of psychological tricks which might lead them away from the choice of PEB. One of the options to construct the desired behavior is default mode condition. Here, the choice is constructed by nudging the person by their environment, which literally demands them to choose the default option. On the example of cycling culture we show that making a cycling environment that is more convenient and comfortable for individuals might provoke the usage of bicycles even if the person does not care about the environmental factor of the transport.

(Yibo Cao) As mentioned above, the choice of travel mode is not just a subjective choice of individuals. Weather can affect cycling behavior, proving that a more convenient cycling infrastructure may help cyclists to ride more even during inconvenient natural conditions. We also see that there are some existing solutions which can make cycling infrastructure more accessible. This may lead to an increase in cycling popularity. Copenhagen can be used as an example to show that when cycling planning is integrated into urban design, bike usage can remain stable or even increase in popularity. Tilted garbage bins, conspicuous lanes, "green waves" projects, bicycle traffic lights and other infrastructure not only provide convenience for cyclists, but also provide a strong cycling atmosphere, thus encouraging more non-cyclists to join the cycling team.

From the above Copenhagen and similar case studies, we know that the usage of bicycles in different regions is largely inconsistent, and the usage condition of bicycles is affected by climate, geographical location, cycling infrastructure, cycling culture, travel distance, cycling policies and other factors. So despite the success of cycling in Denmark, we have to create an infrastructure taking into consideration local conditions and already existing resources. Engineering solutions already exist, but they should be tailored to each unique situation.

In addition to the above environmental factors such as infrastructure and cycling projects, we should also pay attention to the psychological factors in cycling. People are also facing a decision-making process when choosing which way to travel. In people's daily decision-making, it is difficult to make a decision with rationality only. For example, when one is tired after a busy day, it is difficult for them to make a choice based on the distance, the possible environmental benefits of cycling, and the possible health benefits of cycling. Usually at this point, one just wants to go home and get some rest. At such moments, decisions are often made according to heuristics, such as the availability or anchoring heuristic. The availability heuristic refers to the fact that people tend to make decisions based on whether an event or phenomenon is easy to extract in memory. The anchoring heuristic is when people make an initial estimate or judgment and then adjust their decision based on it, but they do not adjust it very much. In areas where the cycling atmosphere or cycling culture is not strong, bicycles cannot be taken as a preferred mode of transportation in memory, and people often choose cars to travel. It might be because of the imperfect cycling infrastructure and the small number of cyclists that it is difficult for people to use bicycles as "anchors" when choosing travel modes. Therefore, environmental factors are closely related to psychological factors. Except for additional cycling education, improving the cycling environment can promote the popularity of cycling by improving people's familiarity with and acceptance of cycling. To put it simply, when we provide more convenient cycling routes for people, more people might choose to cycle.

In the case study, we can find that open data is crucial for cycling-related research. We need more sensors to collect traffic flow data. More data also means more accurate calculations. By adding sensors and other information receiving devices, urban planners can improve facilities and formulate policies according to the actual conditions of each region.

(Elizaveta Baranova-Parfenova) Thus, we observed that the environment indeed affects behavior of some people. When the default option is environmentally friendly, psychological factors such as eco-anxiety may be reduced. Such action becomes not only an environmental one, but also individuals do not need to put in extra effort to make a choice or find extra resources. This may make it easier to manage eco-anxiety if the majority of the environment promotes sustainable defaults.

Another great consequence of default mode is related to the diffusion of responsibility. The implementation and financing of the sustainable environment demands a sufficient amount of research and discussions of the topic. In many cases such changes would be implemented using the money of taxpayers. Thus, default options would put more responsibility on governments and companies, leading to an easier demand system from society.

To conclude, it is important to note that making the most

environment-friendly option a default one is not the solution to all of the problems. The work with an individual's values and level of knowledge is a crucial step towards a more sustainable society, including demands to the governments and corporations to take the responsibilities for their actions and profits. At the same time, a default mode is a great way to navigate an individual toward the more sustainable choice without reducing the amount of options.

4.1.8 *Links*

[1] - https://www.bikeradar.com/features/routes-and-rides/the-uks-best-and-worst-cycling-cities/ (accessed 19 January 2023)

[2] - https://data.cambridgeshireinsight.org.uk/dataset/cambridgeshire-annual-cycle-counts-2018 (accessed 19 January 2023)

[3] - https://www.timeanddate.com/weather/uk/cambridge/historic (accessed 19 January 2023)

[4] - https://bambutec.eu/concepts/cycle-way-roof-urban-mobility (accessed 19 January 2023)

[5] - https://www.independent.ie/life/travel/travel-news/waterford-greenway-wows-almost-250000-visitors-in-first-year-36413679.html

[6] - https://drawdown.org/solutions/bicycle-infrastructure

[7] - https://ecf.com/resources/cycling-facts-and-figures

[8] - https://cyclingsolutions.info/embassy/danish-cycling-statistics/

[9] - https://www.swarco.com/stories/greenwave-copenhagen-denmark

[10] - https://alex-m-dyer.medium.com/the-inhumanity-of-car-dependency-3616a3258f3b

[11] - https://www.ramseysolutions.com/saving/car-depreciation

[12] - https://urbanfortcollins.com/greatest-demand-on-tax-dollars/

[13] - https://ecf.com/resources/cycling-facts-and-figures

[14] - https://www.researchgate.net/publication/305842065_Increasing_car_dependency_of_children_should_we_be_worried

References

Blondiau, T., Van Zeebroeck, B. and Haubold, H. Economic benefits of increased cycling. *Transportation Research Procedia* 14, 2306–2313 (2016).

Chen, W., Carstensen, T. A., Wang, R., Derrible, S., Rueda, D. R., Nieuwenhuijsen, M. J. and Liu, G. Historical patterns and sustainability implications of worldwide bicycle ownership and use. *Communications Earth & Environment* 3, no. 1, 1–9 (2022).

De Luigi, G., Sevaldsen, P. and Patterson, T. The Little Book of Green Nudges (2020).

Gorissen, K. and Weijters, B. The negative footprint illusion: Perceptual bias in sustainable food consumption. *Journal of Environmental Psychology* 45, 50–65 (2016).

Hamilton, T.L. and Wichman, C.J. Bicycle infrastructure and traffic congestion: Evidence from DC's Capital Bikeshare. *Journal of Environmental Economics and Management* 87, 72–93 (2018).

Houthuijs, D.J.M., Van Beek, A.J., Swart, W.J.R. and Van Kempen, E.E.M.M. Health implication of road, railway and aircraft noise in the European Union: Provisional results based on the 2nd round of noise mapping (2014).

Kratschmann, M. and Dütschke, E. Selling the sun: A critical review of the sustainability of solar energy marketing and advertising in Germany. *Energy Research & Social Science* 73, 101919 (2021).

Lange, F. and Dewitte, S. Measuring pro-environmental behavior: Review and recommendations. *Journal of Environmental Psychology* 63, 92–100 (2019).

Mackay, C.M. and Schmitt, M.T. Do people who feel connected to nature do more to protect it? A meta-analysis. *Journal of Environmental Psychology* 65, 101323 (2019).

Marshall, W.E. and Ferenchak, N.N. Why cities with high bicycling rates are safer for all road users. *Journal of Transport & Health* 13, 100539 (2019).

Pichert, D. and Katsikopoulos, K. V. Green defaults: Information presentation and pro-environmental behaviour. *Journal of environmental psychology* 28, no. 1, 63–73 (2008).

Schubert, C. Green nudges: Do they work? Are they ethical? *Ecological economics* 132, 329–342 (2017).

Stern, P. C. Psychological dimensions of global environmental change. *Annual review of psychology* 43, no. 1, 269–302 (1992).

Wang, Z., Xue, M., Zhao, Y. and Zhang, B. Trade-off between environmental benefits and time costs for public bicycles: An empirical analysis using streaming data in China. *Science of the total environment* 715, 136847 (2020).

张长征. 设计幸福感哥本哈根的自行车文化. *Yi Shu Yu She Ji* 6, 166–169 (2015).

5 Intersectionality between environmental protection and social-economic development imbalance — *A compilation of cultural comparison surveys on transportation, food and menstrual product choices*

Dongxue Zhang
Institute of Medical Psychology,
Ludwig Maximilian University of Munich,
Germany

Ian Joshi
Delft University of Technology,
the Netherlands

Yabo Zheng
School of Psychological and Cognitive
Sciences, Peking University, China

5.1 Abstract

As the trend of sustainability-oriented policies, products, and lifestyle gain popularity, individuals and governments that do not embrace these trends are often seen as under-educated or selfish. However, this may not be the case at all, as often circumstances and extenuating factors heavily dictate lifestyle choices. This paper explores the intersectionality between environmental protection and social-economical development imbalance, intending to inspire solutions that strike a balance between environmental sustainability and enrichment for marginalized and impoverished groups.

5.1.1 *Introduction*

In pursuit of sustainability, many Western countries are seeing a trend of policies, products and lifestyles gaining popularity. With this comes the mentality that the people or governments that do not adopt these trends are not environmentally conscious or simply selfish, and must be educated or even admonished. However, while the more privileged are feeling good about recycling their fast-fashion clothing and in some cases consider themselves missionaries to the less "environmentally sophisticated", many in the under-developed part of the world are struggling to fulfil human needs as basic as clean water and food. While young Berliners are increasingly inclined to embrace trendy diets like veganism that often feature lower protein intakes, school children in rural China are saving the single egg and bottle of milk in their school meal to share with their siblings. While YouTube is flushed with comparison videos on which menstrual product is the most environmentally friendly (wasting several perfectly good unused products in the process), many communities struggle with menstrual sanitation, not to mention the stigma around menstrual hygiene or safe sex products in some cultures rendering reusable products unattainable. For them, access to pre-sanitized, disposable sanitary products would drastically increase their quality of life. A further example on the national scale could be made as developing nations are continuously being criticized for their carbon emissions, while developed nations outsource emissions and essentially have already maxed out carbon emissions to industrialize themselves which sparked the climate crisis in the first place.

This paper does not aim to criticize environmentally conscious mindsets at all, but seeks to bring awareness of intersectionality between environmental protection and social-economical development imbalance into the discourse, in the hope that it may help bridge the gaps of communication between communities with different economic, historical and cultural backgrounds, and in the end, encourage sustainable growth globally as a whole. The paper gathers information on the different situations and needs between developed European countries (e.g. the Netherlands and Germany) and impoverished areas in Asia (e.g. Nepal and some parts of China) in topics such as those mentioned previously, and work to come up with possible solutions that strike a balance between environmental sustainability and enrichment for marginalized and impoverished groups.

To achieve this goal, this paper focuses on three key topics: sustainable transportation, food, and menstrual health. In-depth analyses are performed that delve into the factors influencing decision-making between regions regarding environmental protection and social-economical development imbalance.

5.1.2 *Sustainable transportation*

Sustainable transportation is essential for sustainable development (United Nations 2021). As stated by the United Nations, "vulnerabilities are unevenly distributed across countries and population groups" and, as such, it is crucial to explore intersectionality in transportation between different geographical regions.

The growing interest in a more sustainable transportation system is derived from the grossly negative impact that the transportation industry has on the environment today. In 2018, CAIT reported that transportation accounted for 8 billion tonnes of carbon dioxide emissions (Climate Watch 2019), which comprises 24% of energy-related emissions. It is clear that this industry constitutes a significant portion of greenhouse emissions, meaning immediate transformative change is required to achieve sustainability.

Litman and Burwell divide the impact of transportation on sustainability into three categories: economic, social, and environmental (Litman and Burwell 2006). The following section aims to explore reasons behind transportation choices made by disadvantaged regions compared to developed countries, for example, why might fewer people in Nepal opt for electric vehicles over petrol-fuelled vehicles compared to European citizens? Thereafter, this paper proposes possible solutions to address these choices and steer communities towards a sustainable transportation system. These solutions are intended to inspire further discussion and are by no means concrete plans to address sustainability goals.

5.1.2.1   Methodology

To elicit the causes of intersectionality in transportation, two methods were employed. Firstly, a questionnaire was conducted with the sample respondents originating from Nepal, China, Germany, and the Netherlands to gain a representative understanding of people's choices from regions contrasting greatly in income.

It must be noted that the demographics of this survey was not representative of the entire population of the respective countries, as we could not reach respondents in rural areas at the time of the study. To counteract this, extensive research was conducted via research papers and online resources.

5.1.2.2   Discussion

Upon exploring why people in low-income countries (LICs) make arguably unsustainable transportation choices compared to people in middle-income countries (MICs), a variety of reasons present themselves ranging from the macro- to microeconomic scale.

Government policies
Government policies and policy goals both influence the decisions of individuals when it comes to transportation. It is important to study the sustainability goals of the Netherlands and Germany compared to that of Nepal and China, to understand the impact these goals may or may not have on current policies.

Firstly, the Netherlands has committed to the reduction of greenhouses by 49% by 2039 and 95% by 2050 compared to 1990 levels (Government of the Netherlands 2019). An example of a measure to achieve this is the introduction of the carbon tax, which is set to gradually increase on an annual basis (European Commission 2021). Regarding transportation specifically, all new cars must be zero-emission by 2030 (Netherlands Enterprise Agency 2022). Additionally, the Netherlands is a "unique testing ground for smart mobility solutions" (Government of the Netherlands, n.d.) which demonstrates the commitment to innovating sustainable improvements to the current transportation infrastructure.

Meanwhile, Germany aims to reduce emissions by 65% by 2030 (DW 2022) and become greenhouse gas neutral by 2050 (Appunn *et al.* 2022). As an example, the "Energiewende" is a long-term strategy to shift Germany towards a carbon and nuclear-free energy system by 2050 (Agora Energiewende, n.d.). The "Klimaschutzplan 2050" (Bundesministerium für Wirtschaft und Klimaschutz 2016), or Climate Protection Plan 2050, also lays out a comprehensive strategy for achieving these climate goals, which includes reducing greenhouse gas emissions, improving energy efficiency, and increasing the usage of renewable energy. These climate policies have obvious repercussions on the transportation industry which is an indicator of why Germany may opt for comparatively more sustainable transportation choices.

In comparison, China's goal is to peak its emissions by 2030 (Davidson 2022). It is important to note that it is currently the world's largest carbon emitter, contributing almost a third of the world's greenhouse gasses in 2020. This has raised many questions about whether the current measures are sufficient to comply with global climate goals. The "National Carbon Market System" (Nakano and Kennedy 2021) is one of China's most important initiatives, in which the total amount of carbon dioxide production is capped for certain industries. There are, of course, many transportation-specific goals such as the introduction of low-emission zones, fuel efficiency standards, and electrical vehicle targets, and the promotion of public transport.

Nepal, on the other hand, ambitiously aims for full net-zero emissions by 2045 (Government of Nepal 2021). An important policy is to reduce greenhouse gas emissions in the country, which is an important part of Nepal's Nationally Determined Contributions (NDCs) under the Paris Agreement (Ministry of Population and Environment, Government of Nepal 2016). This reduction may be achieved through multiple methods such as afforestation or transitioning to renewable energy sources. However, Nepal lacks many concrete transportation-specific climate policies due to limited resources, although the industry should be indirectly tackled via general climate goals.

Both the Netherlands and Germany are developed nations with higher per capita incomes facilitating the investment of resources in more ambitious climate policies. While China and Nepal have also both committed to the reduction of emissions and promotion of renewable energies, their policies may be slightly more limited in scope due to economic constraints. Transportation is greatly influenced by these policies and, as such, sustainable transportation choices may not be in the hand of the individual.

Infrastructure
The existing structures within these countries contribute significantly to decisions made by consumers when considering different transportation options. Infrastructure concerning transportation may include but is not limited to, the underpinning framework for public transportation, electric vehicles, or other sustainable transport options. The following section discusses some of the limitations that regions within Nepal and/or China face compared to the more developed nations that are Germany and the Netherlands.

Firstly, a shortage in a country's electricity supply not only is a hindrance to economic growth but also enkindles less sustainable choices using transportation options that run on petrol instead of electricity. "Load shedding" is a process whereby the central power grid deliberately cuts off electricity supply to certain areas for a period in order to prevent power grid failures due to excessive demand. In Nepal, localities may suffer from no electricity for up to 16 hours a day (Shrestha 2011). "Ravaged by an electricity crisis", Nepal's load-shedding makes it next-to-impossible for electric vehicle owners to charge their vehicles without alternate power sources, which increases the incentive to opt for petrol-based options. Electricity-based public transportation is also difficult to integrate when there is a general lack of electricity supply available due to the massive effect on demand this would have.

To facilitate electric vehicle transportation, a well-planning and widespread charging infrastructure must be in place. Both the Netherlands and Germany accommodate the charging of electric vehicles exceptionally well. Germany continues to expand their charging infrastructure through policies and investments, evidenced by the €1.8 billion investment scheme (European Commission 2022) to establish a dense network of high-power charging stations. An example of a Dutch policy supporting electrical vehicle charging includes the introduction of the policy mandating charging stations at company parking lots. While China does have an extensive charging network, Nepal suffers from a lack of charging stations. Especially outside the urbanized capital city, there exists a minimal number of charging stations which are sparsely placed, which makes long-distance travel with electric vehicles highly inconvenient.

Furthermore, due to poor and inadequate road networks, sustainable transportation becomes very difficult to achieve. While China's road network is dense, it "does not provide efficient transport access to a large part of the country" (Ojiro 2003). Similarly, Nepal's famously poor road conditions with limited access to rural areas mean that comparatively sustainable private transportation options such as electric vehicles become unviable. This also means public transportation is not easily accessed in all parts of these countries. When compared to Germany for example, which boasts one of the most advanced road systems in the world, one can hypothesize why certain individuals make alternate transportation choices due to circumstance.

Vehicle maintenance
Arguably, the examples of Germany and the Netherlands in comparison to that of Nepal reveal a correlation between levels of vehicle maintenance and pollution through the emission of greenhouse gasses. Implementing consistent policies of vehicle maintenance on the national level — as seen in Germany and the Netherlands — minimizes emissions of harmful pollutants, since studies (Hascic *et al.* 2009) have shown that poorly maintained vehicles emit significantly higher concentrations of these pollutants.

The negative effect of poor vehicular maintenance is only magnified by poor traffic management. Traffic only prolongs the time in which vehicles emit greenhouse gases and studies have shown that stationary vehicles amplify greenhouse emissions. It has been proven that "an idling engine burns fuel less efficiently and can produce up to twice the emissions of a car that is moving" (Cheshire East Council 2022).

Alternate transportation

In addition, there is a clear relationship between the existence of an extensive public transportation system and the emission of greenhouse gases. As implemented in Germany and the Netherlands, well-connected train networks, tram systems and public buses facilitate both daily commutes and longer journeys. In addition, bike lanes provide a safe and easy alternate transportation method for commuters searching for a more sustainable travel method. Alternatively, in Nepal and regions of China, public transportation becomes scarce or a highly inconvenient hindrance. In Nepal, for example, despite the incorporation of a public bus system, scheduling is vague and vehicular maintenance is poor.

Inertia

Decision inertia and status quo bias continue to play a large part in why people do not switch to more sustainable transportation options. Due to an affinity towards petrol vehicles and unwillingness to change to modern sustainable solutions, vehicular emission levels persist. It can be argued that this is both the case in Nepal and China, in addition to Germany and the Netherlands, of course. However, the poor infrastructure in these nations magnifies this issue. Social media trends and public opinion heavily influence people's attitudes towards more sustainable transportation such as cycling or electric vehicles.

5.1.2.3   Possible solutions

The following section explores possible solutions that Nepal and China may implement in order to both facilitate and incentivize citizens to make sustainable transportation choices. These proposed solutions must only be considered ideas and not concretely thought-out plans which are guaranteed to initiate change. Further analysis is required to elicit the feasibility of these proposals and other more viable and/or innovative solutions may exist in the market today.

Firstly, it is apparent that government policy changes are required in both Nepal and China to influence consumer decisions regarding transportation and lower greenhouse emissions. While ambitious policy goals are promising, these goals must be supported by concrete and detailed plans, in addition to levels of investment that closely reflect this ambition. Sustainability within transportation must be made a priority. High-emission institutions must be closely regulated while corporations working towards sustainable and low-emission transportation must be supported through complementary policies or subsidies, for example.

Nepal and China may benefit from taking inspiration from innovative policies and initiatives implemented in Germany and the Netherlands. Smart mobility solutions (Government of the Netherlands, n.d.) only serve as an example of an innovative solution that the Netherlands has embraced to tackle the environmental issues caused by the transportation industry.

As mentioned previously, consistent vehicle maintenance policies must be implemented on a national level. However, these policies must be complemented by a widespread educational campaign to raise public awareness regarding the importance of complying with such maintenance policies in order to maximize their effectiveness. Awareness campaigns to educate citizens on the impacts of transportation choices will help tackle the issue of decision inertia and status quo bias from a grassroots level. An adjustment of social trends is necessary in order to actuate change by ensuring new policy is complied with.

In addition, public transport should be made widely accessible as a means of both compensating individuals for the more stringent vehicle maintenance rules, which would inevitably have unequal effects on different social groups and encourage people to opt out of owning a private vehicle altogether. Overall, an effective approach to tackling vehicle pollution must be comprehensive and multi-dimensional — it must include change and commitment at the policy level but crucially must centre public involvement through educational initiatives and by providing substantive incentives for change, for example, through an efficient and accessible public transportation system.

Predictive traffic management systems such as that proposed by Nafi *et al.* (2014) using modern machine learning tools can help significantly lessen the impact of idling vehicles. These tools must be supported by stringent enforcement and compliance with road rules, which can be complemented by educational campaigns in driving schools.

Furthermore, in this ever-changing world, there are plenty of emerging innovative technological solutions to reduce greenhouse emissions by the transportation industry. Many of these solutions must be chosen or implemented by individual citizens, but the introduction of them in the local economy supported by expansive marketing campaigns can ensure that individuals are at least aware that they have the option to adopt these more sustainable solutions. For example, the conversion of an over-polluting vehicle into an electric vehicle (Lairenlakpam *et al.* 2018) is a financially burdensome but increasingly attractive solution for consumers who cannot afford to replace older vehicles. Additionally, the Zem car (TU/Ecomotive 2022), an innovative new vehicle invented by students of the Technological University of Eindhoven, is capable of actively cleaning the air using direct air capture technology (DAC) while driving. A key example also includes hydrogen fuel cell vehicles (Habib and Arefin 2022), which are considered one of the most promising technologies for sustainable transportation by boosting energy efficiency and reducing greenhouse gas emissions.

In conclusion, a multitude of possible measures can be taken to help consumers make more sustainable transportation choices. Many of these stem from governmental policy changes to incentivize citizens to strongly consider greenhouse gas emissions when travelling, but these must be supported by educational campaigns. Innovative new technologies can also lower the burden of the transportation industry on the climate.

5.1.3 *Food*

Food, energy and water, these basic resources are the nexus of the world's sustainability development. Especially as today's human population has expanded on a drastic level, food production has taken up a large part of the current world's energy consumption and emission. Statistics suggest that in modern societies, food production accounts for about 26% of global greenhouse gas emissions. And half of the world's habitable land is cultivated for agriculture (Ritchie and Roser 2020). To truly realize an environmental-friendly style of production and living, it is important to come up with creative solutions to promote sustainable food choices.

The UN defines food sustainability as "the idea that something (e.g. agriculture, fishing or even preparation of food) is done in a way that is not wasteful of our natural resources and can be continued without being detrimental to our environment or health". However, this definition is not sufficient to describe food sustainability across different cultures, as the interpretation can be highly context- and culture-specific, and lacks acknowledgement of different traditional diet habits, domestic food production structures, individual preferences, etc. Thus, the general public has their ideas on food sustainability, which often include terms like animal welfare, local farming, organic food production, and greenhouse gas emission. Mainstream belief also seems to equal "natural food" to "sustainable food", but there is no consensus on "natural" either.

Hunger and malnutrition are such serious problems that the UN has stressed that a profound change in the global food and agriculture system is needed to tackle them (Abraham and Pingali 2020). The World Food Program reports that more than 1 in 9 people worldwide — 821 million people — suffer from starvation every day (Udmale *et al.* 2020), while in some parts of the world unhealthy diets high in sugar and oil are leaving millions struggling with health issues.

Moreover, with the advent of globalization in food production, people are having much more choices in food selection. Consumers' knowledge of sustainable food plays a great role in this process. Before globalization, food variety and access were relatively limited

(Hartmann *et al.* 2021). The development of food economies has led to a shift from a supply-driven food system to a demand-driven one, and consumer preferences have become increasingly important in the process (Hueston and McLeod 2012). The resulting globalized food system has replaced traditional, decentralized, small-scale food production. This development has its advantages but also comes with costs. With an ever-expanding variety of food products from all over the world freely available, it becomes necessary and, at the same time, more difficult for consumers to evaluate products in terms of healthiness, environmental friendliness, and social justice. The ability to select foods with a high nutritional value and a small environmental footprint seems to have become a desirable skill nowadays (Willett *et al.* 2019).

As essential as it is, our food system is fragile. Unsustainable production and consumption, as well as inefficient distribution, can expose the whole chain of the industry to potential disruption. Furthermore, at present, climate change and ecological crises are also threatening food security globally. We cannot afford to test the limit of what our planet has to provide in terms of food. For both present and future generations, a sustainable food system is crucial to ensure that society will provide for itself.

Everyone in the world has their share of responsibility for sustainable food. However, not everyone ends up practising a sustainable, nutritious diet. The mentality of their behaviour may have complex and individualized dimensions. To better promote the idea of food sustainability, it is crucial to explore these factors and come up with approaches that raise awareness.

Our research aims to investigate what matters most in people's consideration of food selection and compare the underlying cultural difference between China and Western countries. We come up with a short questionnaire containing questions on people's knowledge about food sustainability and their intention to promote a sustainable diet style. We also include some scenarios to collect individuals' open-end ideas to precisely find out what people care most when they make choices in food selection.

5.1.3.1 Method

The questionnaires were posted on open online platforms for a diverse respondent pool. In China, we posted the Chinese version of the questionnaire on powercx.com, and sent out the link to cohorts from Peking University, with 81 valid responses, of which 52 are female, with their ages ranging from 18 to 42 ($M = 27.57$, $SD = 12.77$). For Western participants, we collected data using Google Sheets with the English version of the questionnaire, with 44 respondents ($M = 32.05$, $SD = 13.13$) from different countries.

5.1.3.2 Measures

Food sustainability questionnaire
The questionnaire consists of 13 questions, using simple vocabulary. Questions 1–3 are demographic characteristics; 4–6 are linear scales for self-assessment of participants' general understanding of food sustainability, 7–9 are multiple-choice knowledge quizzes on this topic (correct answers are supported by current research and data), and the last four are open-ended questions with specific scenarios to examine the factors affecting their selections on diets.

Linear scales
All three questions are framed into scales ranging from 1 to 5 with captions, indicating their agreement on the statement. In this section, we set inquiries on their knowledge assessment, willingness, and belief in promoting food sustainability. When proposing the statement, we strictly adopted item-based approaches to better sample the individual responses to each different question. For example, the question for willingness is "How much are you willing to perform sustainable lifestyles on food choice?" with answer scales [1 - I have no intention of doing so, 5 - I have strong motivation].

Knowledge quizzes
This section intends to roughly reflect on what level common people know about specific concrete actions supporting food sustainability. We posed three simple scenarios on the food industry and diet selection in

the form of multiple-choice quizzes. All questions have current research data backing up the correct answer. Specially, we offer the choice "Do not know" to the response list to filter out random guessing. For example, the first quiz question is "The largest environmental impact in the food sectors is due to...", and the response list is [*A. Production / B. Storage Packaging / C. Transportation (Ship, truck) / D. Do not know*].

Open-ended questions
This section is an explorative investigation of factors influencing people's attitudes and food selection behaviour. We created four specific scenarios typical for students and young professionals and asked the participants what impact their decisions had in these scenarios. Besides the open questions, we also listed some example factors for participants to pick, food sustainability among them, allowing for quantitative analysis. This way, we want to unravel what's most important to the participants so that in the future, more customized and individualized measures can be taken to promote food sustainability.

5.1.3.3   Results

High intention, relatively lower awareness
Table 1 shows the first six questions in our survey, and the results are shown in the form of mean rating score or accuracy. In the Chinese group, we found that the mean rating for their willingness is 3.28, their belief in the good cause of food sustainability is 4.26, and their self-evaluated understanding of food sustainability is 2.09, indicating that the public's attitude towards food sustainability is relatively positive, though understanding on actions and measures is lacking. They tend to take a neutral ground concerning whether to perform a sustainable lifestyle in food selection. This may also be partly due to a lack of basic knowledge — their accuracy for the three quizzes is 29.5%, 65.4%, and 60.3%, in line with their self-assessment on knowledge of the matter. In conclusion, the general public is willing but not sufficiently informed to carry out sustainable food choices.

Table 1. Mean rating and accuracy of the first six questions.

	Understanding	Willingness	Belief	Q1	Q2	Q3
Western	3.23	3.64	3.34	50%	77.3%	77.3%
Chinese	2.01	3.20	4.16	31%	63%	65%

In the Western group, the results illustrate a similar pattern but the Western participants generally show a better understanding and higher level of willingness for food choices.

Factors in food selection
When asked what may influence people's choice of a plant-based diet, the Chinese participants care the most about individual needs like personal health conditions (73.1% of the participants selected this factor) followed by taste preference (47.4%). Choice range and accessibility also matter, with several participants giving feedback on limited available choices for a plant-based diet. As for the Western group, health (70.5%) is closely followed by ecological awareness (63.6%) and popularity (59.1%). Similar results also come up when participants were asked what they care about picking agricultural products. Most Chinese participants also picked more personal factors like nutrition (69.1%) and flavour (66.7%). In Western groups, the biggest influencing factor is seasonality and freshness (86.4%), with price coming second (68.2%).

When asked about potential barriers faced during practices of food sustainability, most Chinese participants think about lack of awareness (69.2%), personal preference (65.4%), and limited options (61.5%). In the Western group, most concerns arise from limited options (60.5%) and economic concerns (58.1%).

Moreover, we set up a very specific scenario where we asked about what changes they'd like to make to deal with plastic pollution from ordering takeout food. For Chinese participants, most people agree to switch to shops using greener packaging (55.6%). As for the Western group, most people are willing to recycle the containers (72.7%) or switch to shops using less packaging (72.7%).

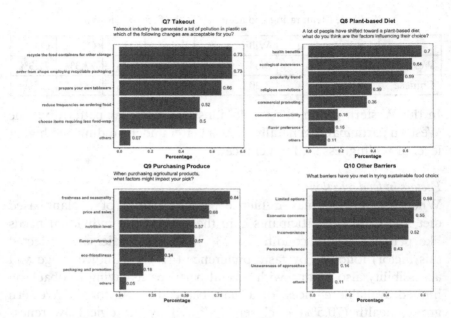

Fig. 1. Factors in choosing food for Western participants.

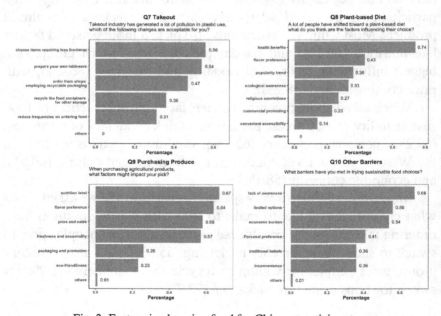

Fig. 2. Factors in choosing food for Chinese participants.

5.1.3.4   Discussion

Our questionnaire aims to take a closer look at certain populations' diet habits and search for better ways to promote and practice food sustainability in everyday life. The results provide us with quantitative data on participants' mentality and from different participant groups, we discovered differences in people's attitudes and behaviours.

Current status for food sustainability
Food sustainability has not exactly been the hottest topic in the environmental protection discourse even among climate change activists. As a result, less information has been passed on to the general public who may have an interest.

In our questionnaire, we find that the general public is not confident with their own understanding of food sustainability despite having a strong intention to adopt a greener diet. This is corroborated by the quantitative data from the quiz questions, which reflect that many people are indeed very unfamiliar with this domain.

Despite the small sample size and simple questionnaire, our results serve as a good illustration of how food sustainability is accepted and practised around the world. We can see that despite years of effort in promoting ecologically friendly ideas, the emphasis on food sustainability is somewhat lacking. We still have quite a long way to go in promoting food sustainability.

Cross-cultural difference
As a culturally diverse group of researchers, we are interested in how cultural background may affect people's choices and behaviour. Overall, both Chinese and Western respondents care for individual needs very much in food selection. However, the Western group is more environmentally minded, in that they are more willing to practice food sustainability in life, more positive about the benefits of promoting food sustainability, and have higher accuracy than Chinese participants in the quiz questions. The results of this survey show that in Chinese participants there are fundamental knowledge gaps regarding the environmental impact of meat production and consumption.

The reasons behind these differences have roots in government policy and social foundations. As the green movements have been going on in the West for decades, the government's regulations concerning eco-friendliness are just starting in China. Recently, due to the pandemic, the main advocacy efforts were shifted towards public health care and sanitary measures, which by nature favour disposable products over reusing.

For the open-ended questions with specific scenarios, we found a similar pattern both in the Chinese group and Western group, with their primary concerns and barriers being individual preference and lacking sufficient options. The slight difference is that the Western participants pay more attention to the cost of a greener diet, as many organic or wholesome food may come at rather high prices. The background here is that organic and green food is not necessarily marketed as "fancy" in China. On the one hand, what the West considers organic and sustainable agriculture is considered rural and lower-end (as contrary to organised and modernised), with China being a traditionally agricultural country with ample cultivatable land, making the price for most agricultural produce relatively economical. On the other hand, Chinese people traditionally take heavy consideration in the nutritious level of their food but less so in their "greenness", so products with greener ecological value are not necessarily expensive.

More effective approaches

As the results have shown, currently the general public still lacks a basic understanding of environmental sustainability measures in the food domain. Much potential benefit of sustainable food choice is overlooked in previous work on environmental protection. Our study shows that participants with different cultural backgrounds lack knowledge of food sustainability information, but are at least self-aware. In the future, researchers and policymakers should put more effort into optimally promoting the notion of food sustainability.

Based on our survey, we can come up with more creative approaches to promote and adopt sustainable diet styles. For instance, we have found out that the most important factors in making decisions are closely related to individual needs like nutrition level and flavour preference. These characteristics should then be emphasised when advocating for greener food choices to sway customers. In fact, a common organic food marketing strategy in China is to stress its health benefits rather than ecological value. Much like how the panda is a flagship species for multi-species habitat protection, health benefits can be the flagship feature for ecologically beneficial foods.

In previous efforts of advocating ecologically friendly measures and ideas, some people are reluctant to open up and oftentimes find the strategies overwhelming, even misunderstanding the goal of sustainability pursuit as political propaganda and turning to the opposite belief. We found these remarks also in our open-ended questions. When we inquired about the barrier to practising a greener lifestyle, some participants gave feedback that mentioned their doubts about the effectiveness of our effort and questioned some strong statements and opinions. If we want to form a firm alliance working together, we need the cooperation of every individual. And that is why we have to find common ground between customers' needs and sustainability goals. From a behaviouristic perspective, getting the general public to act in ways that objectively benefit the environment, even if they subjectively do not mean it, is already significantly mission accomplished. The overall cognition can shift gradually in time towards a conscious understanding of the importance of sustainability.

Limitations

Our study is so far mainly explorative, with severe limitations in sample size and reach. The results in this survey may only apply to given scenarios and participant groups. In the future, if researchers or policy makes aim to further quantitatively measure the knowledge level or attitude of the public, a refined sampling method, and adapted questions should be considered.

5.1.3.5 Conclusion

In this study, we carried out online surveys on people's general attitudes toward food sustainability. Then in closer examination using specific scenarios, we found the most important factors in consumers' decision on food selection were individual needs like flavour preference and health benefits, which outweighs sustainability. To promote healthier notions of consuming food, we must also emphasise what customers want and how sustainable food benefits them, rather than relying solely on their conscience and environmental awareness.

5.1.4 *Menstrual health*

As a self-proclaimed progressive female-bodied person, I have been hearing two voices concerning my menstrual product choices for a while. One is that of feminism, advocating for free choice of menstrual products and improved menstrual hygiene standards. One is that of environmentalism, arguing for more eco-friendly solutions to menstrual hygiene, concerned by the unrecyclable plastic that makes up 90% of menstrual pads (Natracare 2019). Sometimes these voices call for the same thing: to break the taboo, more visibility, and more open discussion around the topic. Sometimes, however, they clash.

In 2019, a post about menstrual cloth pads sparked widespread discussion on Chinese social media. A man wrote about his visit to a museum where he saw the menstrual cloth that Chinese women used in old times and speculated that if more women make use of reusable menstrual products like this instead of disposable ones, it will contribute greatly to preserving the environment. He then mentions that his wife, a Western woman, has been using reusable pads for years. The post was met with all kinds of reactions from Chinese netizens, with some calling him brainwashed by Western ideology, some viewing it as an attack on the progress of women's liberation in China or outright anti-feminist, while others viewing it as progressive and in line with the universal call for environmentalism.

The controversy surrounding this discussion is a microscopic reflection of the role intersectionality plays when ideologies that we may hold as universally are called into practice under different biomes of our global society. I believe that the most common pitfall for a global citizen, policy maker and movement leader, is to take what their cultural upbringing impresses upon them — and in that, almost everything they implicitly learn — as global truth. Even as a member of a marginalized group, one's experience does not represent all, let alone take intersectionality into account and speak for those that may share the same vulnerability as you, yet different on many other fronts.

There are studies on menstrual practices within a region, e.g. France (Parent *et al.* 2022), Taiwan (Tu *et al.* 2021), as well as comparative studies for urban vs rural areas in a given region (e.g. Deo and Ghattargi 2005, Ha and Alam 2022). The intersectionality that I am witnessing here, however, runs deeper. The lack of previous study between China and the West is justified in the huge cultural and societal difference that leaves too much to be considered in a viable comparative questionnaire design: the product available in the market, infrastructure availability, cultural taboo that runs thousands of years deep, bodily anatomy... Yet it is necessary to put China and the World side by side, about a topic that is by nature less discussed than all the other differences we already face and are trying to understand.

Research on the environmental impact of menstrual products almost unanimously points to disposable pads having the biggest negative environmental impact and recommends reusable products for less plastic waste (e.g. Blair *et al.* 2022, Harrison and Tyson 2022). Unlike other disposable plastic products like plastic bags, however, the usage of menstrual products concerns intimately first and foremost the menstruating bodies of their users, thus having everything to do with the diverse bodies and minds of menstruating folks, and everything the society imposes upon them. To address the usage of menstrual products is to address the body autonomy of menstruating folks, and its environmental impact on their choice and practice. It is for this reason we conducted this menstrual practice survey, in the hope that the result can shed some light on how to move forward with menstrual hygiene sustainability.

5.1.4.1   Method

We composed a questionnaire with three sections: demographic info, menstrual product choice and considerations when choosing these products, and influence factors that may lead to a change in product choice. The questionnaire was then imported into Google Forms and distributed online. A Chinese translation of the form was made for Chinese participants on a Chinese survey platform and distributed online.

It should be noted that due to time and resource constraints, the distribution of the questionnaires is done primarily via sharing the questionnaire links through the online social media sphere of the researchers, i.e. by posting in online communities the researchers are, asking friends, family and colleagues to fill it out, etc. The demographic is therefore potentially biased towards well-educated, "progressive" young women and not an accurate sample of the general population. In particular, we lack access to groups on the lower end of the social-economic status scale, whose choice is impacted by this imbalance the most. However, the Chinese population and the English-speaking population surveyed are still comparable, as the general social-cultural-economic development disparity between China and the global West becomes the biggest between-group variable. As such, we will be comparing the results from the Chinese and English versions of the same questionnaire.

5.1.4.2   Results and discussion

Demographics
136 valid responses were collected from the Chinese version of the questionnaire and 97 from the English version. The average age of Chinese respondents is 28.04 (SD = 9.07), and the non-Chinese respondents' age average at 30.56 (SD = 11.11). Of the Chinese respondents, 42.27% currently reside in first-world countries outside China (i.e. Germany, USA, UK etc.), while 90.72% of the English questionnaire respondents reside in first-world countries. Those that do not reside in first-world countries are distributed in Nepal and China.

Product usage
This section is designed to gain insight into the current menstrual practice and mindset from the consumers' side. 97% of Chinese respondents currently use menstrual products regularly. 15% of English respondents used menstrual products regularly but no longer do due to

medical reasons. This disparity likely reflects three things: firstly, the older average age of the English group indicates a higher likelihood of menopause; secondly, the more common usage of menstruation-altering contraception in the West may have contributed to it; and thirdly, perhaps most significantly, one of the online communities the English survey was posted in has a big proportion of transgender people, many of which stopped menstruating due to the medical procedures of gender-reassignment therapy.

When asked about familiarity with different menstrual products, the two groups showed interesting similarities and divergence (Fig. 3). Disposable menstrual pads are the most common primary menstrual product for both groups (40.2% for the English group and 72.1% for Chinese), with it also being the most common secondary menstrual product for the English group (38.1%). The situation with tampons is similar between the two groups as well, with on average 15% of respondents using it as the primary product, although there seems to be a preference for tampons with applicator than without in the Chinese group (24.5% vs 7.2% as primary use and 14.4% vs 12.2% as secondary) and vice versa in the English group (13.40% with applicator vs 16.50% without as primary and 6.18% vs 19.6% as secondary).

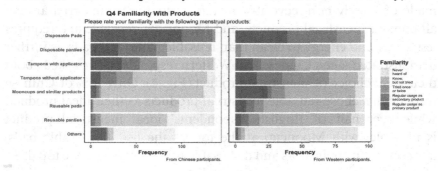

Fig. 3.

Interestingly, the biggest divergences between the two groups are in two products that are perhaps on the two extreme ends of sustainability: disposable panties and Mooncups. Mooncups, reusable silicone cups that work by being inserted into the vagina and sit on the cervix to collect blood and emptied regularly, ranked second highest as the primary menstrual product in the English group with 24.7%, while 92.9% of Chinese respondents have never used it even once. An important finding is that while there is a solid proportion of regular users of Mooncup and an overwhelming amount of people that have heard of it but never tried it, few have tried and then given up, or only use it as a secondary product (3.1% respectively in the English group). This is likely due to the initial cost of reusable products being relatively high and a sunk cost once purchased. Understandably, customers would like to make sure they can commit to it before trying.

Disposable menstrual panties, on the other hand, exhibit the complete opposite pattern. No respondent that answered the English questionnaire has ever tried it as a menstrual product, and 26.8% of them have never even heard of it. On the contrary, every Chinese respondent has heard of disposable panties, or "安心裤", probably due to their wide usage during the pandemic by female medical staff in protective coveralls working long hours, and associated donation efforts organized by feminist charity groups. 23% mark them as their primary menstrual product, and an additional 30.2% use them as secondary.

The contrast of popularity between disposable panties, being made of largely non-recyclable material using more material among all popular menstrual products, and Mooncups, reusable and requires less water and energy to clean than reusable pants and pads, is further demonstrated when Chinese and Western respondents are asked about their ideal choice of menstrual product (Fig. 4). Only 10% of Chinese respondents chose reusable menstrual products as their ideal product, while over half of English respondents' ideal menstrual product is reusable, with Mooncup at the top of the list. Disposable pads, tampons with applicators and disposable panties make up the top three of Chinese respondents' ideal choice list.

On the surface, there seems to be a clear preference towards products that produce less waste in the West than in China. However, when asked about factors that they consider when making their choice on a three-point scale ("not at all" — "to some degree" — "important"), the two groups show very similar concerns (Fig. 5). Over 85% of both groups agree that "ease to use" and "comfort to wear" are important.

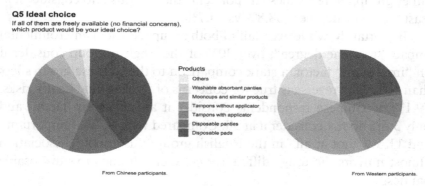

Fig. 4.

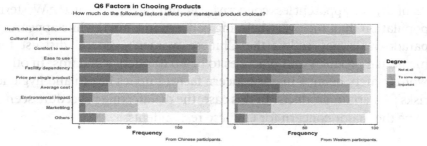

Fig. 5.

Despite differences in public infrastructure like access to public toilets and cleaning supplies, two groups rated similarly on how much they take into account when choosing menstrual products, with around half considering it "important" and around 40% considering it "to some degree". Concerns regarding costs are similar too. Contrary to popular assumption, there is also not as much divergence in how cultural stigma and peer pressure is taken into account in menstrual product choice from the respondent of this survey. Very few people think it is an "important" factor, although there are more Chinese respondents who consider it "to some degree" than English (20.4% vs 7.2%). However, the lack of divergence here is most likely the result of sampling bias and would not reflect that of the general population. Additionally, while marketing is not considered an important factor by either group, more Chinese respondents than English do consider it at least "to some degree" (38.8% vs 24.7%).

Importantly, while over half of both groups consider environmental impact "to some degree", over 30% of the English group consider it an "important" factor, a stark comparison to the Chinese group's less than 10%. The reverse is true for concerns of hygiene and health risks. 79.1% of Chinese respondents consider it an important factor, and only 3.6% do not consider it at all. Compared with 43.3% "important" and 11.3% "not at all" in the English group, this double dissociation almost mirrors the usage difference between Mooncup vs disposable panties.

A few implications can be made here. Firstly, sustainability consciousness seems to be linked to less wasteful product choices; secondly, the apparent less concern for health and hygiene in the Western population than the overwhelming concern in China, while appearing paradoxical, may reflect the Chinese women's general distrust for menstrual products and practices to be up to safety standards. In other words, safety being a major concern factor implies health and hygiene risks for certain products. In this case, the disinfecting procedure seems to be the major concern for Chinese respondents.

Behavioural change influence factors
This section is designed to find out possible entry points to facilitate customer behavioural change from an outside perspective for those who wish to push for more sustainable menstrual product usage, be it by sharing information or making relevant changes as policymakers and manufacturers.

Both groups responded similarly here. The mere disclosure of new information without any material change would already make many people reconsider their product choice, among them health risks ranking the highest in both groups, with instruction tutorial and first-hand experience sharing all welcomed by 40%–75% of the respondents. When it comes to outside changes that would motivate a change of choice in menstrual products, convenience and comfort again ranked top, influencing around 85% of respondents from both groups. Accessibility of products came second, influencing 49% of

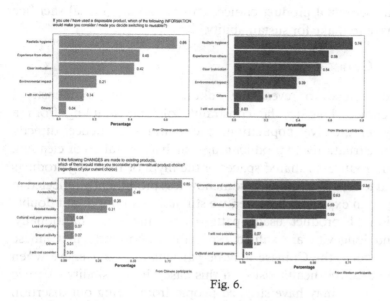

Fig. 6.

Chinese and 63% of Western respondents. Noticeably, price and facility accessibility seem to be able to influence more Western respondents than Chinese ones. Finally, while cultural and peer pressure together with "loss of virginity" hardly impacts the choice of any Western participants, it does impact a small but noticeable number of Chinese respondents (8% and 7%).

These results convey some promising prospects. It shows that most menstrual product users are open to changing their practice upon an exposition of relevant information. Considering the "taboo" status of menstruation in society even today, this speaks to the importance of a louder, unapologetic female voice on menstruation in the public sphere, as well as open, positive and unashamed sharing of information from the manufacturer and scientists. Meanwhile, comfort and convenience remain the core of success for a new menstrual product. While we acknowledge that reusable products are inherently less convenient than disposable ones, more effort should be made to streamline the cleaning process of reusable products, perhaps reaching a point where the inconvenience of cleaning is comparable to that of disposing of used products, in order to make reusable products truly competitive as a good menstrual product choice, rather than a personal sacrifice women have to make for sustainability.

5.1.4.3   General discussion

This study aimed to reveal the cultural difference in menstrual product choice. Due to the constraints of our sampling from a politically progressive population, the cultural influence directly around menstruation and product usage, such as the taboo of cleaning menstrual products in shared spaces or the myth of inserting products would lead to loss of virginity, is somehow less prominently reflected in this study than expected; however, we still found an important double dissociation of product usage pattern and mindset. Traditionally, health and longevity are valued especially importantly in Chinese culture. As such, the Chinese may be particularly risk-aversive when there are perceived health risks. In this case, the possibility of toxic shock syndrome may have stopped people from trying out insertion menstrual products like tampons and Mooncup; moreover, while one can at least somewhat trust disposable menstrual product manufacturers to have professional disinfecting equipment and follow national safety standards, having to sanitize reusable menstrual product with what is available at home or even in public restrooms, there is no simple way of knowing how well the product has been disinfected. This uncertainty may be a major factor holding back people that would like to try out reusable products. Improvement of reusable products towards an easy sanitization process and simple way to confirm may go a long way; failing that, open and scientific information assuring people of their health impacts — for example, designing an experiment showcasing germ culture comparison between properly cleaned reusable products and disposable ones might help, too.

Another important factor is convenience. Despite the Western reusable menstrual pads advocate we mentioned at the start of this section, reusable pads and panties are simply not that popular even among the Western population. The cleaning, drying and storage of soiled products while on the move are simply too much to handle for many. And this brings us back to the fundamental question of our research: is the liberation, rights, time, energy, money etc. of menstruating people, who are themselves a disenfranchised and underprivileged population, worth trading off for the environment? And how much impact would the trade-off make, anyway?

We could not find data directly stating the percentage of menstrual waste in household waste, but we could make a comparison. According to a UK study, 28,114 tons of waste is generated annually from menstrual products in the UK (Blair *et al.* 2022). In comparison, 2.5 million metric tons of plastic packaging waste were generated in the UK in 2021, 44% of which is recycled (Tiseo 2023). That is almost 100 times the volume of menstrual waste. Meanwhile, the safety and convenience of disposable products make billions of women's lives easier, so that they can spare time and energy to fight the many other obstacles women face in our society. We should also not forget menstruating people with special needs — some menstruating respondents that suffer from gender dysphoria reported that many menstrual products trigger their gender dysphoria either by heavily gendered packaging, or the act of vaginal insertion. Some have sensitive sensory issues and have trouble with certain products.

Are we saying that sustainable menstrual practice is not worth pursuing? Certainly not. Even 1% is more than nothing. However, the wants and needs of menstruating people are equally important, as it is an injustice to impose even more on half of the already disadvantaged population by simply being who they are. An ideal solution involves enthusiastic advocacy for the environment as well as women's rights. Mooncup, for example, is reported by many regular users to be comfortable and convenient, while also producing near-zero waste. Even from the perspective of liberating women, it is superior to disposable products in many ways. If the product could be made more available in China, more facts and information are shared openly and honestly, and a sanitization technique satisfying Chinese needs could be developed, then it is a win-win for women and the Earth.

5.1.5 *Conclusion*

Sustainability is more than environmental protection. It is about better, fair human lives at no detriment to the Earth. All this intersectionality reminds us, yet again, that sustainability never has a clear-cut one-size-fits-all solution. Pitching one facet of sustainability against another is what stalls our progress; it is up to us, the future leaders, to understand the needs and wants of a diverse population, and reach a solution where we develop together and no one, human or earth, is left behind.

References

Abraham, M. and Pingali, P. Transforming smallholder agriculture to achieve the SDGs. *The Role of Smallholder Farms in Food and Nutrition Security* 173–209 (2020).

Agora Energiewende. (n.d.). *The German Energiewende.* Agora Energiewende. Retrieved February 02, 2023, from https://www.agora-energiewende.de/en/the-energiewende/the-german-energiewende/.

Appunn, K., Eriksen, F. and Wettengel, J. *Germany's Greenhouse Gas Emissions and Energy Transition Targets.* Clean Energy Wire. Retrieved January 31, 2023, from https://www.cleanenergywire.org/factsheets/germanys-greenhouse-gas-emissions-and-climate-targets (2022).

Blair, L. A. G., Bajón-Fernández, Y. and Villa, R. An exploratory study of the impact and potential of menstrual hygiene management waste in the UK. *Cleaner Engineering and Technology* 7, 100435 (2022).

Bundesministerium für Wirtschaft und Klimaschutz. *Klimaschutzplan 2050* (2016).

Cheshire East Council. *Vehicle Idling.* Cheshire East Council. Retrieved February 24, 2023, from https://www.cheshireeast.gov.uk/business/environmental_health/local_air_quality/air-quality-awareness/vehicle-idling.aspx (2022).

Climate Watch. *Data Explorer.* Climate Watch. Retrieved January 21, 2023, from https://www.climatewatchdata.org/data-explorer/historical-emissions?historical-emissions-data-sources=cait&historical-emissions-gases=co2&historical-emissions-regions=All%20 Selected&historical-emissions-sectors=total-including-lucf%2Ctransportation&page=1& (2019).

Davidson, H. *Is China Doing Enough to Combat the Climate Crisis?* The Guardian. Retrieved February 20, 2023, from https://www. theguardian.com/world/2022/nov/11/china-climate-crisis-renewable-energy-goals (2022).

Deo, D. S. and Ghattargi, C. H. Perceptions and practices regarding menstruation: a comparative study in urban and rural adolescent girls. *Indian Journal of Community Medicine* 30, no. 1, 33 (2005).

DW. *Germany Revamps Climate Plan After Missing Targets.* DW. Retrieved February 20, 2023, from https://www.dw.com/ en/germany-ministers-revamp-climate-plan-after-missing-targets/a-62463199 (2022).

European Commission. *Ensuring That Polluters Pay The Netherlands.* 10.2779/723716 (2021).

European Commission. *State Aid: Commission Approves €1.8 Billion German Scheme to Roll Out High Power Charging Infrastructure for Electric Vehicles* (2022).

Government of Nepal. *Nepal's Long-Term Strategy for Net-Zero Emissions* (2021).

Government of the Netherlands. (n.d.). *Mobility, Public Transport and Road Safety.* Government.nl. Retrieved February 20, 2023, from https://www.government.nl/topics/ mobility-public-transport-and-road-safety.

Government of the Netherlands. *Climate Policy.* Government.nl. Retrieved January 21, 2023, from https://www.government.nl/ topics/climate-change/climate-policy (2019).

Ha, M. A. T. and Alam, M. Z. Menstrual hygiene management practice among adolescent girls: An urban-rural comparative study in Rajshahi division, Bangladesh. *BMC Women's Health* 22, no. 1, 86 (2022).

Habib, M. S. and Arefin, P. Adoption of Hydrogen Fuel Cell Vehicles and its Prospects for the Future. *Oriental Journal of Chemistry* (2022).

Harrison, M. E. and Tyson, N. Menstruation: Environmental impact and need for global health equity. *International Journal of Gynecology & Obstetrics* (2022).

Hartmann, C., Lazzarini, G., Funk, A. and Siegrist, M. Measuring consumers' knowledge of the environmental impact of foods. *Appetite* 167, 105622 (2021).

Hascic, I., de Vries, F., Johnstone, N. and Medhi, N. Effects of Environmental Policy on the Type of Innovation: The Case of Automotive Emission-control Technologies. *OECD Journal: Economic Studies* (2009).

Hueston, W. and McLeod, A. Overview of the global food system: changes over time/space and lessons for future food security. In *Improving food safety through a one health approach: workshop summary. National Academies Press (US), Washington, DC* (2012).

Lairenlakpam, R., Thakre, G., Kumar, P. and Gupta, P. Electric Conversion of a Polluting Gasoline Vehicle into an Electric Vehicle and its Performance and Drive Cycle Analysis. 10.1109/PEDES.2018.8707824 (2018).

Litman, T. and Burwell, D. Issues in Sustainable Transportation. *Int. J. Global Environmental Issues* 6(4). 10.1504/IJGENVI.2006.010889 (2006).

Ministry of Population and Environment, Government of Nepal. *Intended Nationally Determined Contributions (INDC)* (2016).

Nafi, N. S., Khan, R. H., Khan, J. Y. and Gregory, M. A Predictive Road Traffic Management System Based on Vehicular Ad-Hoc Network. *2014 Australasian Telecommunication Networks and Applications Conference (ATNAC)*. 10.1109/ATNAC.2014.7020887 (2014).

Nakano, J. and Kennedy, S. *China's New National Carbon Trading Market: Between Promise and Pessimism.* Center for Strategic and International Studies. Retrieved February 21, 2023, from https://www.csis.org/analysis/chinas-new-national-carbon-trading-market-between-promise-and-pessimism.

Natracare. "How Much Plastic Is in Period Pads?" *Natracare*, 9 Feb. 2021, https://www.natracare.com/blog/how-much-plastic-in-period-pads/.

Netherlands Enterprise Agency. *Electric Transport in the Netherlands.* Netherlands Enterprise Agency. Retrieved February 20, 2023, from https://english.rvo.nl/information/electric-transport (2022).

Ojiro, M. Private Sector Participation in the Road Sector in China. *Transport and Communications Bulletin for Asia and the Pacific* (2003).

Parent, C., Tetu, C., Barbe, C., Bonneau, S., Gabriel, R., Graesslin, O. and Raimond, E. Menstrual hygiene products: A practice evaluation. *Journal of Gynecology Obstetrics and Human Reproduction* 51, no. 1, 102261 (2022).

Ritchie, H. and Roser, M. Environmental impacts of food production. *Our World in Data* (2020).

Shrestha, R. S. Electricity Crisis (Load Shedding) in Nepal, Its Manifestations and Ramifications. *Hydro Nepal Journal of Water Energy and Environment* (6). 10.3126/hn.v6i0.4187 (2011).

Tiseo, I. Topic: Plastic waste in the UK. *Statista*. Retrieved March 6, 2023, from https://www.statista.com/topics/4918/plastic-waste-in-the-united-kingdom-uk/#topicOverview (2023).

Tu, J. C., Lo, T. Y. and Lai, Y. T. Women's cognition and attitude with Eco-friendly menstrual products by consumer lifestyle. *International Journal of Environmental Research and Public Health* 18, no. 11, 5534 (2021).

TU/Ecomotive. *Zem - Cleaning the Air while Driving.* TU/ecomotive. Retrieved February 25, 2023, from https://www.tuecomotive.nl/our-family/zem/ (2022).

Udmale, P., Pal, I., Szabo, S., Pramanik, M. and Large, A. Global food security in the context of COVID-19: A scenario-based exploratory analysis. *Progress in Disaster Science* 7, 100120 (2020).

United Nations. *Sustainable Transport, Sustainable Development* [Integrancy Report for Second Global Sustainable Transport Conference] (2021).

Willett, W., Rockström, J., Loken, B., Springmann, M., Lang, T., Vermeulen, S. and Murray, C. J. Food in the Anthropocene: the EAT-Lancet Commission on healthy diets from sustainable food systems. *The Lancet* 393, no. 10170, 447–492 (2019).

6   Promoting pro-environmental behaviour based on the psychology of perception — *Can the metaverse be a potential solution?*

Lea Skapetze, Chen Zhao and Xuanzi Yin
Institute of Medical Psychology,
Ludwig Maximilian University, Munich,
Germany

David von Brueck
Zeppelin University, Friedrichshafen, Germany

Ezgi Önkü
Eskişehir Osmangazi University,
Eskişehir, Turkey

6.1   Abstract

Climate change and its effects on planet Earth have received increasing public attention within the past two decades. The problems caused by climate-damaging human activities have become innumerable and seem to put the living generations under great pressure to mitigate the progress of global warming. Nevertheless, a majority of people struggle with implementing climate-friendly behavior in their daily lives. In this paper, we try to answer the question of why it may be difficult to implement climate-friendly behavior in our daily lives by taking an environmental-psychological perspective. Furthermore, we discuss the challenges and opportunities of a Metaverse-like virtual 3D world and analyze its potential to promote climate-friendly behavior in the real world. We make suggestions of how different key players could use a Metaverse to help solve some of the pressing problems going hand in hand with climate change.

6.1.1 *Introduction*

Why is it that people all over the world struggle with caring enough about their environment? The most obvious answer to this question might be that most of us have not been directly affected by the changes in climate — yet. We cannot perceive climate change, we cannot see it, smell it, hear it, or touch it in a sufficiently immediate way. It is nowhere near our daily lives, we think of it to be a threat still far away from becoming reality. Too far away to make us reflect on potentially impactful and individual decisions. We do not think of the polar ice melting when putting the plastic waste in the wrong bin, when taking the car and not the bus, or when not considering a meat-free alternative for our next meal. Of course, there has already been a huge shift towards a more responsible and sustainable way of living in Western countries within the past two decades. But we are still far from having implemented climate change awareness into the lives of most people. So, what can we do to spread the idea of a more conscious and balanced way of living our lives, in line with our environment? The Metaverse is a term that has already been described in 1995 (Stephenson 1995) for the first time but only recently got real-world attraction when Facebook became Meta in October 2021 (Rodriguez 2021). The idea of creating a parallel, virtual universe from scratch is appealing and exciting to some, frightening and repulsive to others. In all events, it bears many opportunities as well as risks in how humanity can form its way of living as a global, interconnected society on planet Earth. The concept of a Metaverse allows us not only to rethink education, work, and communication but also gives us the chance to make new, otherwise inaccessible experiences. Associated technologies, such as Virtual Reality (VR) glasses and similar virtual reality experience systems, can bring things that would be too far away in the real world, into our immediate vicinity. In this paper, we are approaching the question of why climate-friendly action is such a great challenge for humanity from the side of perceptual psychology. On the basis of these insights we then try to elaborate on whether the Metaverse might help us solve these big questions of tomorrow or will it rather divide us more and more in a world where we would actually need to find our way back into balance.

6.1.2   *The perception of climate change*

Climate change has a global effect in every region through extreme weather events like heat waves, floods, and droughts and through the damage to ecosystems and species (e.g. Barnett *et al.* 2004, Jacob and Winner 2009, Møller *et al.* 2010). It also has a great impact on human health in physical and psychological ways (Haines and Patz 2004, Padhy *et al.* 2015) Even though there are countless articles published every year about the climate crisis and it is an often addressed topic in the media, not everyone is aware of the problems going hand in hand with man-made climate change. According to a survey by Pew Research Center conducted in 2018 in 26 different countries that asks people the question if global climate change is a major or minor or no threat to their country, most people said global climate change was a major threat to their nation. In countries like South Korea, Greece, Spain, and France, most people expressed very high levels of concern about the climate crisis. On the other hand, more than 50 percent of Israel and Russia and 39 percent of the United States people considered climate change to be a minor threat or not a threat at all (Rosenberg 2019). Even though the percentages of this survey can be positively interpreted, the results cannot be counted as globally representative, especially when we take less developed countries into consideration. According to another survey from Bangladesh, which is one of the most vulnerable countries to climate change, 45.8 percent had not heard of the climate crisis at all (Kabir *et al.* 2016). The rates are similar in countries like Nigeria (Asekun-Olarinmoye *et al.* 2014) and Nepal (Banstola *et al.* 2013).

There are several factors that are correlated with the knowledge and risk perception of the climate crisis for people — structural, psychological, social, and cultural factors like education level, age, gender, income rate, occupation, language, experience, and many others

(Ajuang *et al.* 2016, Calculli *et al.* 2021, T. M. Lee *et al.* 2015, Mustafa *et al.* 2019). As can be seen in the study from Bangladesh, education level and availability of schools in neighborhoods are significantly associated with climate change knowledge (Kabir *et al.* 2016). Sex and age also impact people's knowledge on climate change. It has been seen that men compared to women and older people compared to younger people are more skeptical about the climate crisis and its effects (Whitmarsh 2011). The knowledge, perception, and attitude of people have great importance in bringing communal solutions to this problem since the climate crisis risk perception of people is closely related to their adaptive behavior and mitigation action (Simon *et al.* 2022).

Even though we know about the climate crisis and the anticipated damage that goes along with it for many years now, the concern and mitigation action might not be enough to achieve the goals in order to limit global warming. We are asking the question: Why is it so hard to take action? One of the major obstacles that prevent people from taking action on climate change is the fact that it is difficult to be directly perceived in daily life. Since changes in climate patterns can only be observed over a longer period of time, climate change itself remains an abstract phenomenon that can only be really understood through the help of statistics and mathematical models (Anderson *et al.* 2016, Khan *et al.* 2020, Szulejko *et al.* 2017). It is perceived by people as a distant problem in time and space (Leiserowitz 2005). Many people have not had a personal experience of the change and its potential consequences yet, which is an important note when acknowledging previous findings that personal experience with climate change does predict intentions to act (Broomell *et al.* 2015). In a study from the United States (Myers *et al.* 2013) participants were asked to declare their personal experience with climate change and confirm their belief in climate change's existence. Both times that the survey was conducted (in 2008 and 2011) they found that the belief certainty was higher among participants who said they had personally experienced the effects of global warming than those who said they did not (Myers *et al.* 2013). According to another survey, temperature changes occurring within a shorter time frame, for example, transient changes on a daily basis, can influence the public's opinions and decisions about global climate change. Participants who think that today is warmer than usual believe more in global warming and show greater concern about it. The participants who thought the day was colder than usual were less concerned about climate change and donated less money to a global-warming charity when compared to the group who thought the day was warmer than usual (Li *et al.* 2011). Furthermore, a direct experience of a climate-related natural catastrophe can affect the perception and the motivation for behavioral changes about climate change. A study from the United Kingdom (Spence *et al.* 2011) found that people who experienced a flood expressed significantly more concern about climate change and were less uncertain that climate change exists when compared to people who did not report a flood experience in the past. It has not only been shown that people with a flooding experience were more concerned about climate change but also believed more in the effects of their actions on global warming trends (Spence *et al.* 2011).

In the light of these and similar studies, we have concluded that personal experience and place-based education strategies hold considerable potential in order to help individuals to understand the damage caused by climate change and adapt their behavior in order to mitigate these effects.

6.1.3 *An introduction to the concept of a metaverse*

The term Metaverse was first used in 1995 (Stephenson 1995) a concept based on virtual reality technology which refers to a persistent 3D universe that can be reached via the Internet. It is supposed to connect users in virtual space represented by their avatars and give them the opportunity to interact and socialize by playing games, working, sharing information, etc. (Dionisio *et al.* 2013). To this date Metaverse-like video games exist where it is possible to play a variety of games with virtual partners and that allow an interactive design of the user interface at the same time (Rospigliosi 2022). Different technological devices like VR headsets, eye movement trackers, and voice controllers are being used to access these games and give the feeling of physical presence to the users. This way the Metaverse provides the possibility to directly perceive features of environments that would otherwise be out of reach. At this point, we would like to ask the question: Can this virtual reality technology be used as a tool to educate people about climate change, to help them perceive its effects and to therefore promote climate-friendly behavior?

6.1.3.1   The metaverse — looking at both sides of the coin

The metaverse has many unique characteristics. Bojic (2022) has discussed eight central claims of the metaverse in his study, including richer self-expression, better immersion, better socializing, symmetric relation of physical and virtual spaces, independent markets, better user interfaces, and high demands for regulation and governance. Some of them are the upsides of the metaverse when we use it to combat global climate change, protect the environment, and achieve sustainable human development. Previous studies and experiments have already demonstrated at least four upsides.

First, the metaverse offers an immersive experience. People report higher levels of presence when immersed in virtual environments (Guadagno *et al.* 2007). Davis and his colleagues (2009) have defined the metaverse as an immersive three-dimensional virtual world where people can break the limitations of the natural world and interact as avatars with each other and with software agents. Lo and Tsai (2022) proposed that the metaverse creates a region that mimics the natural world and has functions including stereoscopic vision, hearing, touching, and experiencing. Communicating environmental changes and scenarios to stakeholders and decision-makers is challenging because people outside academia prefer a familiar scene to textured meshes. Immersive environments help transfer knowledge (Gross *et al.* 2022). Rillig and his colleagues (2022) assumed that a metaverse is an excellent tool for more effective communication and sharing of biodiversity. Immersion in virtual worlds is an unprecedented method to share the wonders of biodiversity because they are challenging to reach geographically, like in remote regions like evolutionary trajectories or microorganism worlds.

All this could increase appreciation of and willingness to protect nature. Some psychological theories support this idea. Multimedia-learning theory suggests that the combination of pictures and words will have a better learning result than the results of using picture or word separately (Lo and Tsai 2022). The construal level theory explains that the psychological distance, how far we perceive an object or event, influences how concretely or abstractly the thing is mentally described (Trope and Liberman 2010). The long psychological distance of climate change might be rooted in the fact that one can rarely see environmental problems and horrible results in the real world (Brügger *et al.* 2015, Loy and Spence 2020). Some experimental results have supported this idea as well. Lo and Tsai (2022) have proved that participants who received related instruction improved in learning environmental conservation concepts in an experiment. Plechatá and her colleagues (Plechatá *et al.* 2022) have demonstrated in their study that one week's virtual reality intervention significantly reduces dietary carbon consumption because of increased intentions, self-efficacy, risk perception, and emotional reactions. VR interventions might promote pro-environmental behavior in VR users. All of the above experiments last only for a limited period. When the Metaverse becomes a life habit, it may be hypothesized to have a lasting dynamic influence on the users. Therefore, the metaverse could be a suitable platform for environmental protection by increasing human ecological awareness.

Second, the metaverse is interactive and can provide a good communication platform. The metaverse is a world where humans can communicate, collaborate, and socialize using intelligent devices. Humans are the center of the metaverse (Jaber 2022). The metaverse connects many users or devices, greatly facilitating user communication (Sun *et al.* 2022). In addition, it can make virtual scenario representations to experts, policymakers, and the general public (Gross *et al.* 2022). Therefore, it can strengthen global cooperation on sustainable development through communications between hubs using the latest metaverse services (Umar 2022).

Third, people can use the metaverse to predict future development and make plans. Huang (2022) pointed out that in the metaverse, the time attribution is retroactive, among the past, present, and future. Therefore, geographical environment, policy guidance, economic development, humanities, and other factors should also be important when considering the global situation. The metaverse can build a decision-making model for future environmental development with deep learning algorithms and combining the influences of multiple data sources. The numerous studies in many domains such as architecture (Huang *et al.* 2022), accounting and auditing (Al-Gnbri 2022), urban planning (Hudson-Smith and Shakeri 2022), healthcare industry (C. W. Lee 2022), marketing and branding (Nalbant and Aydin 2023) have already proved the predictive and planning abilities of the metaverse. Umar (2022) pointed out that we can use the metaverse to establish a complicated model for climate change prediction, such as changes in ocean waves, weather conditions, and earthquakes.

Fourth, the metaverse is a significant advance in science and technology, which can more accurately and efficiently carry out activities to respond to environmental changes. Sun (2022) pointed out that The Metaverse is an emerging technology of the future, which has its roots in big data, AI (artificial intelligence), VR (Virtual Reality), AR (Augmented Reality), MR (mixed Reality), and other technologies. The metaverse will encourage all related development in software and hardware industries, including big data, cloud computing, blockchain, cybersecurity, latency-sensitive networks, virtual Reality, and augmented Reality. Computing power is fundamental to the metaverse. At least ten times of current computing power is required. Cloud storage and computing technologies can make data-intensive and computing-intensive tasks more efficient and orderly (Tang *et al.* 2021). 5G and 6G technology is the current technological outcome of combining communication and storage with data. The transmission capacity of 6G can be 100 times higher than that of 5G, up to 1 Tbps (Zhou *et al.* 2020). With the support of the above-advanced techniques, we can solve climate change problems more effectively and accurately.

Some downsides of the metaverse emerge during its development. Many researchers and studies noticed that the metaverse could not solve the climate change problem efficiently and effectively in the short term. First, the metaverse needs to be more convenient and comfortable. Pyo (2022) found that the difficulty of accessing and sharing files on the metaverse causes inconveniences. Cheng and his colleagues (2022) supposed that the current metaverse has problems in both scalability and accessibility. Second, the metaverse is only affordable to some people in some countries. Umar (2022) pointed out that the metaverse requires expensive equipment, a significant drawback for developing nations. Mystakidis (2022) agreed that the high equipment cost was a barrier to mass adoption in the long run. The high price has already become a natural barrier for those least developed or developing countries to get involved in the metaverse.

In addition to that, it creates new problems as well. First, the metaverse increases energy consumption. Rillig and his colleagues (2022) have noticed that powering the metaverse will have enormous energy demands. Suppose humans cannot find alternative energy sources. It will increase greenhouse emissions, contribute to climate change, and adversely affect ecosystems and biodiversity. It contradicts our original intention of using the metaverse to solve the problem of climate change. Second, the metaverse causes both physical and psychological health-related issues. Mystakidis (2022) pointed out that on the physical level, the attention distraction of users in location-based AR applications has led to harmful accidents. Information overload is a psychological challenge as well. Pellas (2021) listed motion sickness, including nausea, dizziness, and head and neck fatigue, as some negative results. Slater (2020) cited some psychological problems caused by the metaverse, such as addiction, social isolation, and abstinence from actual, physical life. Third, the metaverse will arouse security and privacy risks unavoidably. Mystakidis (2022) assumed that the additional data layer would emerge as a possible cyber security threat. Many privacy problems arise from data gathered by Virtual Reality technology, such as retina scans, biometric data in face geometry, fingerprints, and voice prints. The above reasons caused security and privacy vulnerability. In short, many new problems appeared and deteriorated during the metaverse development. Jaber (2022) cited social engineering attacks, ransomware attacks, network credential theft, and identity theft as examples.

6.1.3.2   An outlook to the future — examples of how a metaverse might help to promote climate-friendly action

The preceding reflections lead us to think about practical implementations of a Metaverse-like concept and ways that it could have a positive impact on the promotion of climate-friendly behavior. We want to approach this thought experiment in a non-valuing way, and it remains to be discussed whether the potential benefits a Metaverse could have on the environment outweigh the downsides that come along with these changes.

If we want to change environmental behavior, we need to identify the most important actors, and we need suitable mechanisms to make a difference. There are various models of predictors of environmental behavior that try to combine external and internal factors that also take into account the individuality of a person, their sense of responsibility, potential cultural and family influences on values, attitudes, awareness, emotion, and priorities (Kollmuss and Agyeman 2002). However, in our thought experiment on the impact of a Metaverse on pro-environmental behavior, we decided to focus on external factors primarily. Therefore, we divided our suggestions and ideas into three main groups that might have considerable effects on the climate-related behavior of individuals: Education, Economy, and Activism. Schools and educational institutions have a big impact on the development of environmental awareness and pupils' action competence for pro-environmental behavior (Cincera and Krajhanzl 2013). The will to act in an environment-friendly way is also dependent on the level of economic development in a country (Mikuła *et al.* 2021) and can be largely influenced by the presence of methods used by climate activism that is taking place in the surroundings of an individual person (Fisher and Nasrin 2021). Furthermore, we also acknowledge the influence of governmental institutions and politics on the climate-related behavior of individuals but extensive analysis of these influences is out of the scope of this paper and might need to be discussed by further research.

Within the different categories of external influence, a further differentiation between bottom-up influences and top-down influences can be made. Bottom-up influences are basically driven by the people of a certain population or country that may not hold a specific leadership position in this society. The outcome of their actions is limited by their lack of consensus and the missing ability to implement their collective actions into policies and binding rules. Top-down influences are induced by the institutions, organizations, and instruments of our civilizations. The outcome of their actions is limited by their lack of cooperation, their different interests, and the failure in setting up policy mechanisms as well as guaranteeing their compliance on a global level to eliminate the competition mindset between the actors (Ostrom 1990).

6.1.4 *Education: Stimulate new ways of thinking*

Education is vital, and its forever partner is learning. Education should never be the approach to tame children, but the dogma of life and learning. Education in the general sense is to spread correct facts and better ideas to people of any age, and even though the information is new to them, making them accept, embrace, and eventually act accordingly to the knowledge should be the goal of education. Better education should serve not only kids but also adults. In top-down regulations, we consider the shortcomings of current school education and the gap between the scientific circle and the public. In schools, there might be courses about environmental pressure and climate change. Indeed, environmental education is an essential strategy, and it helps to establish a positive attitude and increase awareness about the environment in students (Louv 2005). Even though field research and practical implementation are more effective than in-class teaching (Feszterová 2015), it seems that it is still hard to bring long-lasting habits into daily behaviors. For adults, finding channels to learn about edge-cutting scientific research is not easy, even if they have an interest, energy is a limitation. Especially in a fast digital age, it is getting harder to ask people to read long papers in concentration on leisure time. So, the key point of the top-down regulations is to make scientific knowledge easier to be accessed and more interesting to be absorbed. Some feasible good mediums can be videos, games, talks, and methods of "gamification" that might raise pupils' attention and awareness (Zhang *et al.* 2022).

The Metaverse might be able to take these processes to the next level. Within a virtual reality classroom and 3D online universities, it would be possible to bring together teachers and students from all over the world. Students and pupils would have the opportunity to get to know peers from different countries, they could share their thoughts on environmental topics, and collaborate on projects in order to save the planet. A 3D interface would allow a much more immediate interaction, the direct exchange of ideas, and even nonverbal communication. People with different social, cultural, economic, and professional backgrounds would have a chance to connect quickly and easily but highly qualitatively. Within the past years of the Covid-19 pandemic we have seen many universities and schools switching to virtual classrooms and ways of teaching — now could be the chance to use this development and take it to the next level. This could be creating highly flexible educational programs and systems fitting individual interests and talents on the one hand and addressing urgent challenges humanity is facing at the moment and will have to deal with within the upcoming decades. The entire "universe" of knowledge might be accessible to every citizen of the world in the long term, and factors limiting access to education today, such as age, wealth, ethnicity, and geographical borders, could be drastically reduced by creating and establishing universities and schools in a virtual Metaverse. Scientists with different backgrounds might be able to align themselves virtually and collaborate on projects such as the design of different "climate models" in a virtual world where the effects of different climate models on the environment could be measured. Students could be studying the predictions of different climate model scenarios in 3D. This might improve their understanding of environmental processes and lead them to more pro-environmental behavior in the real world.

6.1.5 *Economy: Driving innovative ideas and solutions*

By promoting better education for everyone, we hope to motivate the active ones to learn more about the climate crisis and act correspondingly.

While at the same time, there will definitely be a portion of the public who has neither motivation nor the feeling of responsibility to mend their behavioral routines. This is where the macro environment should play its role. The seemingly free citizens are caged by the structure of society, and their choices are actually pre-defined by the leading giants in any industry. If a small change in the economic system influences a habit of many individuals, then in the public there will be a significant wave of change. For example, just imagine if all the sales departments decide not to provide plastic bags anymore (Bosse *et al.* 2019). Surely the example is unrealistic, but the key point is that decision-makers have to conceive and simulate policies that make it easier and cheaper to be ecologically responsible. This requires foresights and planning in the long run development because it is hard to make a sudden turn in well-established life patterns (Ostrom 1990). In the Autumn school, Professor Elke Pahl-Weber mentioned in her lecture that it is really difficult to persuade the residents to participate in her project, which is to update the energy system in their houses to reduce electric consumption. So people need some kind of incentives for active participation (Pahl-Weber 2017).

But what if we take it a step further? Modern technologies like virtual worlds and Blockchains give us the opportunity to transfer our whole energy market management from the real world to virtual spaces. This concept is called "smart grid energy commons" and offers a bottom-up solution for greener energy. The concept of so-called "smart grids" (regional energy communities) proposes the complete digitization of the energy market and decentralization of energy production. Large parts of the energy grids and management are already being organized digitally and automatically. On the basis of data on local electricity consumption (e.g. trends in metering peaks at different times of the day depending on consumers) and on the type of energy generation (e.g. photovoltaics only works during the day or wind power only when the weather is suitable), for example, predictions can be made about the different electricity requirements of the smart grids at different times and in this way energy can be divided between smart grids that are currently producing in excess and smart grids that are currently having a payload peak have transferred (Acosta *et al.* 2018). The smart grids correspond to the individual self-organized energy communities that are part of a virtually organized network of countless smart grids and therefore do not necessarily have to be geographically linked. Energy communities could stretch thousands of miles because energy is known to travel at the speed of light and energy consumption and price can be billed digitally. With blockchain networks, this could be completely automated and fraud-proof (Melville *et al.* 2017). Studies show that decentralized communities manage their energy production and consumption more efficiently and sustainably than centralized forms of organization (Jenny *et al.* 2007). This behavior results from pure utilitarianism. What good would it do for the indigenous people of the Amazon to make a profit from logging if their livelihoods are being destroyed as a result? Global corporations, on the other hand, can benefit from deforestation and slash-and-burn without having to suffer direct negative consequences, since they do not operate in their own ecosystem and are therefore not related (Brondizio *et al.* 2009). One could also ask how a village community benefits from a coal-fired power station that is right on their doorstep and is harmful to air quality. Hardly anyone would accept that for a lower electricity price. However, if this power plant is far away, yes. The separation of production and consumption places reduces the sense of responsibility for the consequences of energy production (Melville *et al.* 2017). In a nutshell, those interconnected smart grids could be understood as an energy internet or social media platform for managing a decentralized energy market while supporting sustainable production (Giotitsas *et al.* 2020). In addition, the metaverse allows us to experience any smart grid on the market virtually. In this way, how electricity is produced can also be viewed beyond the borders of one's community, thus ensuring transparent and sustainable energy procurement. In this way, the Metaverse could help create more sustainable and even more efficient energy markets which also helps the Metaverse itself get greener, regarding its enormous energy consumption as described above. Of course, we also need some internationally regulated top-down regulations so that, for example, the global North does not only consist of green smart grids and smart grids in the Global South continue to supply us with dirty energy (Acosta *et al.* 2018).

For this reason, it would be helpful if there were a central platform for the energy internet whose guidelines are based, for example, on the Sustainable Development Goals, including binding laws and regulations for charging violations supervised by, for example, the UN (*THE 17 GOALS | Sustainable Development*, o. J.).

However, if the city and architectural constructions were well designed and planned ahead by some top-down regulations, ideally the green choices could be the better choices for the residents themselves. The same logic can be applied to many other aspects, including economics in general. So in the top-down direction, active and cooperative simulation and planning for future development make it possible to optimize the utilization of resources. At the same time, it makes green habits and choices the default life pattern for everyone.

When thinking about the reasons why private companies could invest in the Metaverse, their primary source of interest might not be to promote climate-friendly behavior. But as more and more time and money will be spent in the Metaverse, the shift from real-world to virtual consumption could still greatly impact the environment and natural resources. There are various ideas that came to our mind when thinking about possible ways of how this could look for different industries. As an example, we would like to mention the fashion and textile industries. A wide range of fashion labels have already been engaged in creating virtual clothes, bags, accessories, and even perfumes (*How the Metaverse Can Revolutionize the Fashion Industry* 2022). This might seem trivial at the first glance, but in a future where the average consumption in the Metaverse increases, it might be very beneficial for the environment to prefer shopping for the avatar over buying textiles with poor ecological footprints in the real world.

6.1.6 *Activism: Communicating change in "perception-friendly" ways*

Paradoxically, the Internet has made a very wide range of knowledge and information accessible to everyone, but at the same time users are getting an increasingly targeted flow of information that is determined by adaptive algorithms designed to recommend content related to the topics we like (Rhodes 2022), and communication in quality is even rare than before. This phenomenon is called echo room: when showing a specific interest, you will be trapped in the swirl of it. Statistics show that the echo-room strategy is good for companies to stick the users to their sites (Yusuf *et al.* 2014), but as a consequence, our scope is getting limited, and it becomes harder to learn out of the comfort zone. As a significant result, this trend gradually creates a natural gap between groups, including those who follow environmental topics versus those who do not. Although the internet is free and transparent, there are actually invisible walls. To let knowledge, opinions, and common sense penetrate the walls, takes not only patience and skills, but also attempts at some new approach (Rhodes 2022).

Climate activism relies on attracting the attention of the public. It aims to raise awareness within society for the environment, its reciprocal relationship with humanity, and the potential of change they see in individual behavior as well as large-scale economic and political decisions. Activism, therefore, relies on communicating these ideas in a way that they stick in the heads and minds of the public. Communicating properly is crucial to the success of every campaign promoting pro-environmental behavior.

A Metaverse could allow climate activists to improve their strategies and methods of communication. Organizing voluntary environmental activities would be easier in a virtual space without physical limitations. For example, virtual fundraising has been gaining importance, especially within the past years of the Covid-19 pandemic when it was nearly impossible to host big charity events. An online gala dinner, auction, ceremony, tour, and virtual talks have become routinely used tools to raise money for environmental and humanitarian projects. Not only big organizations can take advantage of the connectivity of the Metaverse to advocate for the public, but also in the bottom-up direction, the Metaverse may amplify the influence of folk organizations as it would also be easier for ordinary people to create their base and raise the voice.

Another way we could imagine the Metaverse to be of use to climate activists is by connecting people in industrialized countries with people in need whose lives have been affected by climate change. Greenpeace and Fridays For Future might want to approach individuals in a Metaverse and show them what life for victims of climate change looks like. This could lead to an increase in awareness of environmental catastrophes and the need to act accordingly in order to prevent them.

6.1.7 *Conclusion*

Albert Einstein once famously stated: "We have to learn to think in a new way" (SB 1955), indicating that in order to solve a problem we cannot approach it in the same way of thinking as we had when creating it. By analyzing the potential use of a Metaverse to raise people's awareness of climate change and its effects on the environment, we tried to put this intent into practice. At the same time, we know of the innumerable limitations going along with our arguments and discussion. Many questions could not be addressed, let alone answered in this text. Further research has to be done on the quality of perception a Metaverse can actually provide and whether the way we can perceive our surroundings through a Metaverse would be sufficient to impact people's behavior in the real world. Especially as this technology is in its infancy to the date we are writing this text, it is very likely that the "Meta-world" noticeably differs from the "real world" in various aspects and features. Therefore, one cannot guarantee that experiences made in one world would impact the behavior of the same person in the other world. Also, when thinking about the ways external influences could have an impact on climate change-related behavior of humanity, it still needs to be answered whether a Metaverse that is provided by a private company or a public government, rather supports the promotion of top-down or bottom-up processes. Geopolitical forces that might play a big role in the equal or unequal distribution of access to the Metaverse have not been addressed in this paper by any means and need to be considered in further analysis. Especially in its early stages, access to new technologies is restricted by individual limitations, such as wealth, place of birth, gender and ethnicity. It is precisely those factors that divide humanity and that we are proposing to overcome by implementing the Metaverse. This potential can only be used in a positive way if it is handled with the required amount of responsibility and if the justice of opportunities is ensured.

References

Acosta, C., Ortega, M., Bunsen, T., Koirala, B. and Ghorbani, A. Facilitating Energy Transition through Energy Commons: An Application of Socio-Ecological Systems Framework for Integrated Community Energy Systems. Sustainability, 10(2), 366 (2018). https://doi.org/10.3390/su10020366.

Ajuang, C. O., Abuom, P. O., Bosire, E. K., Dida, G. O. and Anyona, D. N. Determinants of climate change awareness level in upper Nyakach Division, Kisumu County, Kenya. SpringerPlus 5, no. 1, 1015 (2016). https://doi.org/10.1186/s40064-016-2699-y.

Al-Gnbri, M. K. Accounting and Auditing in the Metaverse World from a Virtual Reality Perspective: A Future Research. Journal of Metaverse 2, no. 1, Art. 1 (2022).

Anderson, T. R., Hawkins, E. and Jones, P. D. CO2, the greenhouse effect and global warming: From the pioneering work of Arrhenius and Callendar to today's Earth System Models. Endeavour 40, no. 3, 178–187 (2016). https://doi.org/10.1016/j.endeavour.2016.07.002.

Asekun-Olarinmoye, E., Bamidele, J. O., Odu, O. O., Olugbenga-Bello, A. I., Abodunrin, O. L., Adebimpe, W. O., Oladele, E. A., Adeomi, A. A., Adeoye, O. A. and Ojofeitimi, E. O. Public perception of climate change and its impact on health and environment in rural southwestern Nigeria. Research and Reports in Tropical Medicine 5, 1–10 (2014). https://doi.org/10.2147/RRTM.S53984.

Banstola, A., Chhetri, M. R., Schneider, R. M., Stebbing, M. and Banstola, A. Knowledge related to climate change and willingness to act for adaptation and mitigation practices in rural Nepal. Vietnam Journal of Public Health 1, no. 1 (2013). https://uwe-repository.worktribe.com/output/3248010/knowledge-related-to-climate-change-and-willingness-to-act-for-adaptation-and-mitigation-practices-in-rural-nepal.

Barnett, T., Malone, R., Pennell, W., Stammer, D., Semtner, B. and Washington, W. The Effects of Climate Change on Water Resources in the West: Introduction and Overview. Climatic Change 62, no. 1, 1–11 (2004). https://doi.org/10.1023/B:CLIM.0000013695.21726.b8.

Bosse, J., Buch, F., Häublein, E. and Socher, S. Klimawende von unten—Wie wir durch direkte Demokratie die Klimapolitik in die Hand nehmen. https://www.klimawende.org/handbuch/ (2019).

Brondizio, E. S., Ostrom, E. and Young, O. R. Connectivity and the Governance of Multilevel Social-Ecological Systems: The Role of Social Capital. Annual Review of Environment and Resources 34, no. 1, 253–278 (2009). https://doi.org/10.1146/annurev.environ.020708.100707.

Broomell, S. B., Budescu, D. V. and Por, H.-H. Personal experience with climate change predicts intentions to act. Global Environmental Change 32, 67–73 (2015). https://doi.org/10.1016/j.gloenvcha.2015.03.001.

Brügger, A., Dessai, S., Devine-Wright, P., Morton, T. A. and Pidgeon, N. F. Psychological responses to the proximity of climate change. Nature Climate Change 5, no. 12, Art. 12 (2015). https://doi.org/10.1038/nclimate2760.

Calculli, C., D'Uggento, A. M., Labarile, A. and Ribecco, N. Evaluating people's awareness about climate changes and environmental issues: A case study. Journal of Cleaner Production 324, 129244 (2021). https://doi.org/10.1016/j.jclepro.2021.129244.

Cheng, R., Wu, N., Chen, S. and Han, B. Will Metaverse be NextG Internet? Vision, Hype, and Reality (arXiv:2201.12894). arXiv. https://doi.org/10.48550/arXiv.2201.12894 (2022).

Cincera, J. and Krajhanzl, J. Eco-Schools: What factors influence pupils' action competence for pro-environmental behaviour? Journal of Cleaner Production 61, 117–121 (2013). https://doi.org/10.1016/j.jclepro.2013.06.030.

Davis, A., Murphy, J., Owens, D., Khazanchi, D. and Zigurs, I. Avatars, People, and Virtual Worlds: Foundations for Research in Metaverses. Journal of the Association for Information Systems, 10, no. 2 (2009). https://doi.org/10.17705/1jais.00183.

Dionisio, J. D. N., III, W. G. B. and Gilbert, R. 3D Virtual worlds and the metaverse: Current status and future possibilities. ACM Computing Surveys, 45, no. 3, 34:1–34:38 (2013). https://doi.org/10.1145/2480741.2480751.

Fisher, D. R. and Nasrin, S. Climate activism and its effects. WIREs Climate Change, 12, no. 1, e683 (2021). https://doi.org/10.1002/wcc.683.

Giotitsas, C., Nardelli, P. H. J., Kostakis, V. and Narayanan, A. From private to public governance: The case for reconfiguring energy systems as a commons. Energy Research & Social Science, 70, 101737 (2020). https://doi.org/10.1016/j.erss.2020.101737.

Gross, F., Petersen, L., Wallmeier, C. and Karstens, S. Coastscapes for a Metaverse: From height maps to virtual vegetated environments [Preprint]. In Review. https://doi.org/10.21203/rs.3.rs-2225992/v1 (2022).

Guadagno, R. E., Blascovich, J., Bailenson, J. N. and Mccall, C. Virtual Humans and Persuasion: The Effects of Agency and Behavioral Realism. Media Psychology, 10, no. 1, 1–22 (2007). https://doi.org/10.1080/15213260701300865.

Haines, A. and Patz, J. A. Health Effects of Climate Change. JAMA, 291, no. 1, 99–103 (2004). https://doi.org/10.1001/jama.291.1.99.

How the Metaverse can revolutionize the fashion industry. Cointelegraph. https://cointelegraph.com/news/how-the-metaverse-can-revolutionize-the-fashion-industry (2022).

Huang, H., Zeng, X., Zhao, L., Qiu, C., Wu, H. and Fan, L. Fusion of Building Information Modeling and Blockchain for Metaverse: A Survey. IEEE Open Journal of the Computer Society, 3, 195–207 (2022). https://doi.org/10.1109/OJCS.2022.3206494.

Hudson-Smith, A. and Shakeri, M. The Future's Not What It Used To Be: Urban Wormholes, Simulation, Participation, and Planning in the Metaverse. Urban Planning, 7, no. 2, 214–217 (2022). https://doi.org/10.17645/up.v7i2.5893.

Jaber, T. A. Security Risks of the Metaverse World. International Journal of Interactive Mobile Technologies, 16, no. 13, 4–14 (2022). https://doi.org/10.3991/ijim.v16i13.33187.

Jacob, D. J. and Winner, D. A. Effect of climate change on air quality. Atmospheric Environment, 43, no. 1, 51–63 (2009). https://doi.org/10.1016/j.atmosenv.2008.09.051.

Jenny, A., Hechavarria Fuentes, F. and Mosler, H.-J. Psychological Factors Determining Individual Compliance with Rules for Common Pool Resource Management: The Case of a Cuban Community Sharing a Solar Energy System. Human Ecology, 35 no. 2, 239–250 (2007). https://doi.org/10.1007/s10745-006-9053-x.

Kabir, M. I., Rahman, M. B., Smith, W., Lusha, M. A. F., Azim, S. and Milton, A. H. Knowledge and perception about climate change and human health: Findings from a baseline survey among vulnerable communities in Bangladesh. BMC Public Health, 16, no. 1, 266 (2016). https://doi.org/10.1186/s12889-016-2930-3.

Khan, M. J. U., Islam, A. K. M. S., Bala, S. K. and Islam, G. M. T. Changes in climate extremes over Bangladesh at 1.5°C, 2°C, and 4°C of global warming with high-resolution regional climate modeling. Theoretical and Applied Climatology, 140, no. 3, 1451–1466 (2020). https://doi.org/10.1007/s00704-020-03164-w.

Kollmuss, A. and Agyeman, J. Mind the Gap: Why do people act environmentally and what are the barriers to pro-environmental behavior? Environmental Education Research, 8, no. 3, 239–260 (2002). https://doi.org/10.1080/13504620220145401.

Lee, C. W. Application of Metaverse Service to Healthcare Industry: A Strategic Perspective. International Journal of Environmental Research and Public Health, 19, no. 20, Art. 20 (2022). https://doi.org/10.3390/ijerph192013038.

Lee, T. M., Markowitz, E. M., Howe, P. D., Ko, C.-Y. and Leiserowitz, A. A. Predictors of public climate change awareness and risk perception around the world. Nature Climate Change, 5, no. 11, Art. 11 (2015). https://doi.org/10.1038/nclimate2728.

Leiserowitz, A. A. American Risk Perceptions: Is Climate Change Dangerous? Risk Analysis, 25, no. 6, 1433–1442 (2005). https://doi.org/10.1111/j.1540-6261.2005.00690.x.

Li, Y., Johnson, E. J. and Zaval, L. Local Warming: Daily Temperature Change Influences Belief in Global Warming. Psychological Science, 22, no. 4, 454–459 (2011). https://doi.org/10.1177/0956797611400913.

Lo, S.-C. and Tsai, H.-H. Design of 3D Virtual Reality in the Metaverse for Environmental Conservation Education Based on Cognitive Theory. Sensors, 22, no. 21, Art. 21 (2022). https://doi.org/10.3390/s22218329.

Loy, L. S. and Spence, A. Reducing, and bridging, the psychological distance of climate change. Journal of Environmental Psychology, 67, 101388 (2020). https://doi.org/10.1016/j.jenvp.2020.101388.

Melville, E., Christie, I., Burningham, K., Way, C. and Hampshire, P. The electric commons: A qualitative study of community accountability. Energy Policy, 106, 12–21 (2017). https://doi.org/10.1016/j.enpol.2017.03.035.

Mikuła, A., Raczkowska, M. and Utzig, M. Pro-Environmental Behaviour in the European Union Countries. Energies, 14, no. 18, Art. 18 (2021). https://doi.org/10.3390/en14185689.

Møller, A. P., Fiedler, W. and Berthold, P. Effects of Climate Change on Birds. OUP Oxford (2010).

Mustafa, G., Latif, I. A., Bashir, M. K., Shamsudin, M. N. and Daud, W. M. N. W. Determinants of farmers' awareness of climate change. Applied Environmental Education & Communication, 18, no. 3, 219–233 (2019). https://doi.org/10.1080/1533015X.2018.1454358.

Myers, T. A., Maibach, E. W., Roser-Renouf, C., Akerlof, K. and Leiserowitz, A. A. The relationship between personal experience and belief in the reality of global warming. Nature Climate Change, 3, no. 4, Art. 4 (2013). https://doi.org/10.1038/nclimate1754.

Mystakidis, S. Metaverse. Encyclopedia, 2, no. 1, Art. 1 (2022). https://doi.org/10.3390/encyclopedia2010031.

Nalbant, K. G. and Aydin, S. Development and Transformation in Digital Marketing and Branding with Artificial Intelligence and Digital Technologies Dynamics in the Metaverse Universe. Journal of Metaverse, 3, no. 1, Art. 1 (2023). https://doi.org/10.57019/jmv.1148015.

Ostrom, E. Governing the Commons: The Evolution of Institutions for Collective Action. Cambridge University Press (1990). https://doi.org/10.1017/CBO9780511807763.

Padhy, S. K., Sarkar, S., Panigrahi, M. and Paul, S. Mental health effects of climate change. Indian Journal of Occupational and Environmental Medicine, 19, no. 1, 3–7 (2015). https://doi.org/10.4103/0019-5278.156997.

Pahl-Weber, E. elke pahl weber tu berlin urban design thinking a new approach to urban cooperation by Anton Shynkaruk—Issuu. https://issuu.com/antonshynkaruk/docs/6.2-elke-pahl-weber-tu-berlin-urban (2017).

Pellas, N., Mystakidis, S. and Kazanidis, I. Immersive Virtual Reality in K-12 and Higher Education: A systematic review of the last decade scientific literature. Virtual Reality, 25, no. 3, 835–861 (2021). https://doi.org/10.1007/s10055-020-00489-9.

Plechatá, A., Morton, T., Perez-Cueto, F. J. A. and Makransky, G. Virtual Reality Intervention Reduces Dietary Footprint: Implications for Environmental Communication in the Metaverse. PsyArXiv. https://doi.org/10.31234/osf.io/3ta8d (2022).

Pyo, K.-H. Comparison of University Students' Perceptions of Interaction in English Classes on Two Online Platforms: Gather. town vs. Zoom. 언어학 연구, 65, 287–312 (2022). https://doi.org/10.17002/sil.65.202210.287.

Rillig, M. C., Gould, K. A., Maeder, M., Kim, S. W., Dueñas, J. F., Pinek, L., Lehmann, A. and Bielcik, M. Opportunities and Risks of the "Metaverse" For Biodiversity and the Environment. Environmental Science & Technology, 56, no. 8, 4721–4723 (2022). https://doi.org/10.1021/acs.est.2c01562.

Rodriguez, S. Facebook changes company name to Meta. CNBC. https://www.cnbc.com/2021/10/28/facebook-changes-company-name-to-meta.html (2021).

Rosenberg, S. Climate Change Still Seen as the Top Global Threat, but Cyberattacks a Rising Concern. Pew Research Center's Global Attitudes Project. https://www.pewresearch.org/global/2019/02/10/climate-change-still-seen-as-the-top-global-threat-but-cyberattacks-a-rising-concern/ (2019).

Rospigliosi, P. 'asher'. Metaverse or Simulacra? Roblox, Minecraft, Meta and the turn to virtual reality for education, socialisation and work. Interactive Learning Environments, 30, no. 1, 1–3 (2022). https://doi.org/10.1080/10494820.2022.2022899.

SB. Statement: The Russell-Einstein Manifesto. Pugwash Conferences on Science and World Affairs. https://pugwash.org/1955/07/09/statement-manifesto/ (1955).

Simon, P. D., Pakingan, K. A. and Aruta, J. J. B. R. Measurement of climate change anxiety and its mediating effect between experience of climate change and mitigation actions of Filipino youth. Educational and Developmental Psychologist, 39, no. 1, 17–27 (2022). https://doi.org/10.1080/20590776.2022.2037390.

Slater, M., Gonzalez-Liencres, C., Haggard, P., Vinkers, C., Gregory-Clarke, R., Jelley, S., Watson, Z., Breen, G., Schwarz, R., Steptoe, W., Szostak, D., Halan, S., Fox, D. and Silver, J. The Ethics of Realism in Virtual and Augmented Reality. Frontiers in Virtual Reality, 1. https://www.frontiersin.org/articles/10.3389/frvir.2020.00001 (2020).

Spence, A., Poortinga, W., Butler, C. and Pidgeon, N. F. Perceptions of climate change and willingness to save energy related to flood experience. Nature Climate Change, 1, no. 1, Art. 1 (2011). https://doi.org/10.1038/nclimate1059.

Stephenson, N. Snow Crash: Roman (J. Körber, Übers.; 2nd edition). Blanvalet Taschenbuch Verlag (1995).

Sun, J., Gan, W., Chen, Z., Li, J. and Yu, P. S. Big Data Meets Metaverse: A Survey (arXiv:2210.16282). arXiv. https://doi.org/10.48550/arXiv.2210.16282 (2022).

Szulejko, J. E., Kumar, P., Deep, A. and Kim, K.-H. Global warming projections to 2100 using simple CO_2 greenhouse gas modeling and comments on CO_2 climate sensitivity factor. Atmospheric Pollution Research, 8, no. 1, 136–140 (2017). https://doi.org/10.1016/j.apr.2016.08.002.

Tang, X., Cao, C., Wang, Y., Zhang, S., Liu, Y., Li, M. and He, T. Computing power network: The architecture of convergence of computing and networking towards 6G requirement. China Communications, 18, no. 2, 175–185 (2021). https://doi.org/10.23919/JCC.2021.02.011.

THE 17 GOALS | Sustainable Development. (o. J.). Abgerufen 19. Januar 2023, von https://sdgs.un.org/goals.

Trope, Y. and Liberman, N. Construal-level theory of psychological distance. Psychological Review, 117, 440–463 (2010). https://doi.org/10.1037/a0018963.

Umar, A. Metaverse for UN SDGs – An Exploratory Study. Science-Policy Brief for the Multistakeholder Forum on Science, Technology and Innovation for the SDGs (2022).

Whitmarsh, L. Scepticism and uncertainty about climate change: Dimensions, determinants and change over time. Global Environmental Change, 21, no. 2, 690–700 (2011). https://doi.org/10.1016/j.gloenvcha.2011.01.016.

Zhou, Y., Liu, L., Wang, L., Hui, N., Cui, X., Wu, J., Peng, Y., Qi, Y. and Xing, C. Service-aware 6G: An intelligent and open network based on the convergence of communication, computing and caching. Digital Communications and Networks, 6, no. 3, 253–260 (2020). https://doi.org/10.1016/j.dcan.2020.05.003.

Index

A

activism 120, 126
adaptation 19
addiction 119
affective empathy 52
afforestation 81
agriculture 20, 87, 94
altruism 52
analytic thinking 41
animal welfare 87
anonymity 31
anxiety 40, 48, 50, 51
architecture 32
artificial intelligence 45, 118
atmosphere 62, 65
attention 48, 119, 126
attitude 114
augmented reality 118
authority 46, 47
automobile 66
autonomy 41, 49
avatars 115, 116, 125
awareness 88, 112
ayahuasca 16, 17, 23

B

balanced sustainability 7
behavior 39, 46, 51, 63, 73, 116, 120, 127
bicycle 49, 62, 64, 65, 66, 72
bicycle usage 43

big data 118
biodiversity 1, 8, 17, 20, 25, 26, 29, 30, 31, 42, 116, 119
blockchain 124
body autonomy 97
bottom-up 120, 127
built environment 3, 4, 7, 8, 10

C

capitalism 29
carbon consumption 117
carbon dioxide 62, 79, 81
carbon emission 67, 78
carbon footprint 49, 50
carbon tax 80
case study 68
causality 46
Chinese culture 104
circular agriculture 31
circulation 6
citizen 84, 86
city dweller 10, 11
city satellites 7
climate activism 126
climate change 1, 3, 9, 29, 42, 43, 47, 48, 54, 61, 62, 88, 93, 111, 112, 113, 116, 118, 126, 127
climate crisis 2, 17, 18, 19, 20, 33, 39, 50, 51, 78, 113
climate model 122
climate policies 82

Printed in the United States
by Baker & Taylor Publisher Services